STUDY GUIDE

Robert E. Nunley • George W. Ulbrick • Bernard O. Williams

University of Kansas   *Johnson County*   *University of Kansas*
Community College

An Introduction to
Human Geography

The Cultural Landscape

EIGHTH EDITION

James M. Rubenstein

PEARSON

Prentice
Hall

Upper Saddle River, NJ 07458

Editor-in-Chief, Science: John Challice
Executive Editor: Dan Kaveney
Associate Editor: Amanda Griffith
Vice President of Production & Manufacturing: David W. Riccardi
Executive Managing Editor: Kathleen Schiaparelli
Assistant Managing Editor: Becca Richter
Production Editor: Rhonda Aversa
Supplement Cover Manager: Paul Gourhan
Supplement Cover Designer: Joanne Alexandris
Manufacturing Buyer: Ilene Kahn
Cover Photo Credit: Skip Nall/Getty Images/Photodisc

© 2005 Pearson Education, Inc.
Pearson Prentice Hall
Pearson Education, Inc.
Upper Saddle River, NJ 07458

Pearson Prentice Hall® is a trademark of Pearson Education, Inc.

The author and publisher of this book have used their best efforts in preparing this book. These efforts include the development, research, and testing of the theories and programs to determine their effectiveness. The author and publisher make no warranty of any kind, expressed or implied, with regard to these programs or the documentation contained in this book. The author and publisher shall not be liable in any event for incidental or consequential damages in connection with, or arising out of, the furnishing, performance, or use of these programs.

Printed in the United States of America

10 9 8 7 6 5 4

ISBN 0-13-142948-5

Pearson Education Ltd., *London*
Pearson Education Australia Pty. Ltd., *Sydney*
Pearson Education Singapore, Pte. Ltd.
Pearson Education North Asia Ltd., *Hong Kong*
Pearson Education Canada, Inc., *Toronto*
Pearson Educación de Mexico, S.A. de C.V.
Pearson Education—Japan, *Tokyo*
Pearson Education Malaysia, Pte. Ltd.

CONTENTS

Preface

The present study guide was produced by the three authors, aided by many students and faculty from the Geography Department of the University of Kansas. Dr. Williams prepared the final draft, and assumes full responsibility for any errors or shortcomings. A web site supplements the text and study guide and is available at http://www.prenhall.com/rubenstein.

It is suggested that this study guide be used as a framework to outline the chapters for a full grasp of concepts and as a means to integrate the material with course notes. The terms at the end of the chapters will be helpful for review.

The present study guide is designed quite differently from those for previous editions. Instead of supplementing the main text with additional materials, the present design leaves that function to the web site on the now ubiquitous Internet. The basic design goal of the present study guide is to provide a uniformly distilled version of the text. The goal is to facilitate recall for anyone who previously read the entire text. It is not assumed that beginning students will find the Study Guide useful without referring to the main text.

An additional planned use of the present study guide is to help students who want to learn to read geographic literature in English. We are offering (at no cost) a machine readable version of the study guide to professionals (in the Geography Department of any university) willing to translate the entire study guide into their native language and make their translation available to us at no cost to us for distribution to others at no charge. The plan is to make it possible for non-English speaking scholars to study inexpensively the translation of the study guide and compare it with the English version of the study guide, and then be able to move to the main text and read it with the assistance of only a dictionary. Conversely, English-speaking students who want to read geographic literature in the language of the translation could use the study guide as a first effort to read introductory geographic concepts in that language.

Comments, inquiries, and suggestions from students and professors about any aspect of the study guide are most welcome. E-mail: nunley@ku.edu and/or berneyw@ku.edu.

(66045, March 29, 2004)

Chapter 1. Thinking Geographically

Contemporary geography is the scientific study of the location of people and activities across Earth, and the reasons for their distribution. The ancient Greek scholar Eratosthenes coined the word geography from two Greek words, *geo* meaning *earth* and *graphy* meaning *to write*. Geography asks two simple questions: *where* and *why*.

Geography is divided broadly into two categories—*human* geography and *physical* geography. Human geography studies where and why human activities are located as they are. Physical geography studies where and why natural forces occur as they do. This book focuses on human geography, but it never forgets Earth's atmosphere, land, water, vegetation, and other living creatures. Because geographers are trained in both social and physical sciences, they are particularly well equipped to understand interactions between people and their environment.

(3)
Key Issues
1. How do geographers describe where things are?
2. Why is each point on Earth unique?
3. Why are different places similar?

(4)
Geographers observe that people are being pulled in opposite directions by two factors: globalization and local diversity. Tensions between the simultaneous geographic trends of globalization and local diversity underlie many of the world's problems that geographers study, such as political conflicts, economic uncertainty, and pollution of the environment.

This book . . . concentrates on two main features of human behavior: culture and economy. Distinctive geographic approaches concentrate on five aspects in thinking about the world: space, place, regions, scale, and connections. These five distinctive ways that geographers think about the world are discussed in detail in this chapter.

(6)
Key Issue 1. How Do Geographers Address *Where* Things Are?
- **Maps**
- **Contemporary Tools**

Geographers think about the arrangements of people and activities found in space and try to understand why those people and activities are distributed across space as they are. Geographers use maps as a method of depicting the distribution of features and as a tool for explaining observed patterns.

(7)
Maps
Geography's most important tool for thinking spatially about the distribution of features across Earth is a map. A **map** is a two-dimensional or flat-scale model of Earth's surface, or a portion of it. For centuries geographers have worked to perfect the science of mapmaking, called **cartography**. A map serves two purposes: a tool for storing reference material and a tool for communicating geographic information. A map is often the best means for depicting

the distribution of human activities or physical features, as well as for thinking about reasons underlying a distribution.

Early Mapmaking
The earliest surviving maps were drawn by Babylonians on clay tablets about 2300 B.C., but mapmaking is undoubtedly even older. Polynesian peoples navigated for thousands of years with three dimensional maps.

Mediterranean sailors and traders made maps of rock formation, islands, and ocean currents as early as 800 B.C.

Aristotle (384–322 B.C.) was the first to demonstrate the earth was spherical. He observed the curved shadow of the earth on the moon during an eclipse and the fact that the visible groups of stars change as one travels north or south. Eratosthenes (276?–194? B.C.), the first person on record to use the word geography, calculated the circumference of the earth and made one of the earliest maps of the known world, correctly dividing Earth into five climatic regions.

Ptolemy (A.D. 100?–170?) wrote an eight-volume *Guide to Geography*, taking advantage of information collected by merchants and soldiers who traveled throughout the Roman Empire.

(Non-European). After Ptolemy little progress in mapmaking or geographic thought was made in Europe for hundreds of years, although geographic inquiry continued outside of Europe. Phei Hsiu (or Fei Xiu), the "father of Chinese cartography," produced an elaborate map of China in A.D. 267. The Muslim geographer al-Idrisi (1100–1165?) prepared a world map and geography text in 1154, building on Ptolemy's work. Ibn-Battutah (1305–1368?) wrote *Rihlah* (Travels) based on three decades of journeys.

(Age of Exploration and Discovery). Geography and mapmaking enjoyed a revival during the Age of Exploration and Discovery. By the seventeenth century, maps accurately displayed the outline of most continents and the position of oceans.

(9)
Map Scale
The **scale** of a map is the relation of a feature's size on a map and its actual size on Earth's surface. Map scale is represented in three ways: a fraction (1/24,000) or ratio (1:24,000), a written statement ("1 inch equals 1 mile"), or a graphic bar scale.

A graphic scale usually consists of a bar line marked to show distances on Earth's surface. The bar line is used by measuring a distance on the map, then reading that distance along the bar line. The appropriate scale for a map depends on the information being portrayed.

(11)
Projection
To communicate geographic concepts effectively through maps, cartographers must design them properly and assure that users know how to read them. Earth's spherical shape poses a challenge for cartographers because drawing Earth on a flat piece of paper unavoidably produces some distortion. The scientific method of transferring location on Earth's surface to a flat map is called projection.

Four types of distortion can result: *shape* can be distorted, *distance* may be increased or decreased, *relative size* may be altered, and *direction* between points can be distorted. Most of the world maps in this book are *equal area projections*.

U.S. Land Ordinance of 1785

In addition to the global system of latitude and longitude, other mathematical indicators of locations are used in different parts of the world. In the United States, the Land Ordinance of 1785 divided much of the country into a system of townships and ranges to facilitate the sale of land to settlers in the West.

In this system, a township is a square 6 miles on each side. Some of the north-south lines separating townships are called principal meridians and some east-west lines are designated base lines. Each township has a number corresponding to its distance north or south of a particular base line. Each township has a second number, known as the range, corresponding to its location east or west of a principal meridian. A township is divided into 36 sections, each of which is 1 mile by 1 mile. Each section is divided into four quarter-sections. A quarter-section, which is 0.5 mile by 0.5 mile, or 160 acres, was the amount of land many western pioneers bought as a homestead.

(12)
Contemporary Tools

Two important technologies that developed during the past quarter century are remote sensing from satellites (to collect data) and geographic information systems (computer programs for manipulating geographic data).

GIS

A **geographic information system (GIS)** is a high-performance computer system that processes geographic data. Each type of information (topography, political boundaries, population density, manufacturing, soil type, earthquake faults, and so on) is stored as an information layer. GIS is most powerful when it is used to combine several layers, to show relations.

Remote Sensing

The acquisition of data about Earth's surface from a satellite orbiting Earth or from other long-distance methods is known as remote sensing. The smallest feature on Earth's surface that can be detected by a sensor is the resolution of the scanner. Some can show an object 1 meter across. Weather satellites take a broader view, looking at several kilometers at a time.

(14)
GPS

The **Global Positioning System (GPS)** is an example of applying new technology to an old human habit: consulting a map to get to a desired destination. The GPS can pinpoint a location using signals from a group of satellites.

(15)
Key Issue 2. Why Is Each Point on Earth Unique?
  * **Place: Unique location of a feature**
  * **Regions: Areas of unique characteristics**

The interplay between the uniqueness of each place and the similarities among places lies at the heart of geographic inquiry into why things are found where they are. Two basic concepts help geographers to explain why every point on Earth is in some ways unique: place and region. The difference between the two concepts is partly a matter of scale: a place is a point, whereas a region is an area.

Humans possess a strong sense of place—that is, a feeling for the features that contribute to the distinctiveness of a particular place, perhaps a hometown or vacation spot.

Place: Unique location of a feature
Geographers identify the location of something in four ways—by place-name, site, situation, and mathematical location.

Place Names
Geographers call the name given to a portion of Earth's surface its **toponym** (literally, place-name). The name of a place may give us a clue about its founders, physical setting, social customs, or political changes. Some place-names derive from features of the physical environment. The name of a place can tell us a lot about the social customs of its early inhabitants. The Board of Geographical Names, operated by the U.S. Geological Survey, was established in the late nineteenth century to be the final arbiter of name on U.S. maps. Places can change names, possibly to commemorate a particular event. After the fall of communism in the early 1990s, names throughout Eastern Europe were changed, in many cases reverting to those used before the Communists had gained power a few decades earlier.

(16)
Site
The second way that geographers describe the location of a place is by **site**, which is the physical character of a place. Important site characteristics include climate, water sources, topography, soil, vegetation, latitude, and elevation. Humans have the ability to modify the characteristics of a site. The central areas of Boston and Tokyo have been expanded through centuries of landfilling in nearby bays.

Situation
Situation is the location of a place relative to other places. Situation is a good way to indicate location for two reasons—finding an unfamiliar place and understanding its importance. Many locations are important because they are accessible to other places.

(18)
Mathematical Location
The location of any place on Earth's surface can be described precisely by meridians and parallels, two sets of imaginary arcs drawn in a grid pattern on Earth's surface. A **meridian** is an arc drawn between the North and South poles. A **parallel** is a circle drawn around the globe parallel to the equator. The location of each meridian is identified on Earth's surface according to a numbering system known as **longitude**. The **prime meridian**, 0° longitude, passes through the Royal Observatory at Greenwich, England. All other meridians have numbers between 0° and 180° east or west of Greenwich.

The numbering system to indicate the location of a parallel is called **latitude**. The equator is 0° latitude, the North Pole is 9° north latitude, and the South Pole is 90° south latitude. The mathematical location of a place can be designated more precisely by dividing each degree into 60 minutes and each minute into 60 seconds.

Determining Longitude. Measuring latitude and longitude is a good example of how geography is partly a natural science and partly a study of human behavior. Latitudes are scientifically derived by Earth's shape and its rotation around the Sun. On the other hand, 0° longitude is a human creation. The 0° longitude runs through Greenwich because England was the world's most powerful country when longitude was first accurately measured and the international agreement was made.

Telling Time from Longitude. Longitude plays an important role in calculating time. Traveling 15° east is the equivalent of traveling one hour forward on the clock, and 15° west is one hour backward.

Earth is divided into 24 standard time zones, one for each hour of the day, so each time zone represents 15° of longitude. Before standard time zones were created, each locality set its own time. The international agreement (in 1884) designated the time at the prime meridian as **Greenwich Mean Time (GMT)** or **Universal Time (UT)**. The eastern United States, which is near 75° west longitude, is five hours earlier than Greenwich Mean Time. When you cross the **International Date Line**, which for the most part follows 180° longitude, you move the clock back 24 hours, or one entire day, if you are heading eastward toward America. You turn the clock ahead 24 hours if you are heading westward toward Asia.

(20)
Regions: Areas of Unique Characteristics
An area of Earth defined by one or more distinctive characteristics is a **region**. A region derives its unified character through the **cultural landscape,** a combination of cultural features such as language and religion, economic features such as agriculture and industry, and physical features such as climate and vegetation.

Cultural Landscape
The contemporary **cultural landscape** approach in geography—sometimes called the **regional studies** approach—was initiated in France. It was later adopted by several American geographers, who argued that each region has its own distinct landscape that results from a unique combination of social relationships and physical processes.

Types of Regions
Geographers most often apply the concept region at one of two scales: either several neighboring countries that share important features, such as those in Latin America, or many localities within a country, such as those in southern California. A particular place can be included in more than one region depending on how the region is defined. Geographers identify three types of regions: formal, functional, and vernacular.

(20)
Formal Region. A formal region, also called a uniform region or a homogeneous region, is an area within which everyone shares in common one or more distinctive characteristics.

Some formal regions are easy to identify, such as countries or local government units. In other kinds of formal regions a characteristic may be predominant rather than universal. For example, the wheat belt of North America also grows other crops. A cautionary step in identifying formal regions is the need to recognize the diversity of cultural, economic, and environmental factors, even while making a generalization.

(22)
Functional Region. A functional region, also called a nodal region, is an area organized around a node or focal point. The region is tied to the central point by transportation or communications systems or by economic or functional associations. An example of a functional region is the circulation area of a newspaper. New technology is breaking down traditional functional regions. Newspapers such as *USA Today* and the *New York Times* are transmitted by satellite to printing machines in various places.

Vernacular Region. A **vernacular region**, or perceptual region, is a place that people believe exists as part of their cultural identity. Such vernacular regions emerge from people's informal perceptions of place, rather than from scientific models. As an example of a vernacular region, Americans frequently refer to the South as a place with environmental, cultural, and economic features perceived to be quite distinct from the rest of the United States.

(23)
Spatial Association
Different conclusions may be reached concerning a regions characteristics depending on scale. For example, death rates vary widely among scales within the United States.

At the national scale, the eastern regions of the United States have higher levels of cancer than the western ones. At the scale of the state of Maryland, the city of Baltimore and counties in the east have higher levels of cancer than the western and suburban counties. At the scale of the city of Baltimore, lower levels of cancer are found in the zip codes on the north side. To explain why regions possess distinctive features, such as a high cancer rate, geographers try to identify cultural, economic, and environmental factors that display similar spatial distributions. Geographers conclude that factors with similar distributions have spatial association.

Regional Integration of Culture
In thinking about *why* each region on Earth is distinctive, geographers refer to **culture**, which is the body of customary beliefs, material traits, and social forms that together constitute the distinct tradition of a group of people. Intellectually challenging culture is often distinguished from *popular culture*, such as television programs. *Culture* also refers to small living organisms, such as those found under a microscope or in yogurt. *Agriculture* is a term for growing things on a much larger scale. The origin of the word *culture* is the Latin *cultus*, which means "to care for," which has two very different meanings.

Culture. Some geographers study what people care about (their ideas, beliefs, values, and customs), whereas other geographers emphasize what people take care of (their ways of earning a living and obtaining food, clothing and shelter).

(24)
What People Care About. Especially important cultural values derive from a group's language, religion, and ethnicity. Language is a system of signs, sounds, gestures, and marks

that have meanings understood within a cultural group. Religion is an important cultural value because it is the principle system of attitudes, beliefs, and practices through which people worship in a formal, organized way. Ethnicity encompasses a group's language, religion, and other cultural values, as well as its physical traits, products of common traditions, and heredity.

What People Take Care Of. The second element of culture of interest to geographers is production of material wealth—the food, clothing, and shelter that humans need to survive and thrive. Different cultural groups obtain their wealth in different ways. Geographers divide the world into regions with countries that are more (or relatively) developed economically (abbreviated MDCs), and regions with less developed (or developing) countries (abbreviated LDCs). Agriculture predominates in LDCs, while manufacturing and performing services for wages predominates in MDCs. Some manufacturing is leaving MDCs and relocating in LDCs. Geographers are also interested in the political institutions that protect material artifacts, as well as cultural values. As discussed in Chapter 8, cultural groups in the modern world are increasingly asserting their rights to organize their own affairs at the local scale rather than submit to the control of other cultural groups.

Cultural Ecology: Integrating Culture and Environment
In constructing regions, geographers consider environmental factors as well as cultural. Cultural ecology is the geographic study of human-environment relations. Some nineteenth century geographers argued that human actions were *scientifically caused* by environmental conditions, an approach called **environmental determinism**. Modern geographers reject environmental determinism in favor of **possibilism**, arguing that the physical environment may limit some human actions, but people can adjust to their environment. People choose a course of action among alternatives in the environment, and endow the physical environment with cultural values by treating it as substances for use, a collection of **resources**. For example, the climate of any location influences human activities, especially food production.

(26)
Human and Physical Factors. Human geographers use this cultural ecology, or human-environment, approach to explain many global issues. People can adjust to the capacity of the physical environment by controlling their population growth, adopting new technology, consuming different foods, migrating to new locations, and other actions. A people's level of wealth can also influence its attitudes toward modifying the environment. A rocky hillside is an obstacle to a farmer with a tractor, but an opportunity to a farmer with a hoe. Modern technology has altered the historic relationship between people and the environment.

Physical Processes: Climate. Human geographers need some familiarity with global environmental processes to understand the distribution of human activities.

Climate is the long-term average weather condition at a particular location. Geographers frequently classify climates according to a system developed by German climatologist Vladimir Köppen. The modified Köppen system divides the five main climate regions into several subtypes. The climate of a particular location influences human activities, especially production of the food needed to survive.

(24)
Physical Processes: Vegetation. Plant life covers nearly the entire land surface of Earth. Earth's land vegetation includes four major forms of plant communities, called biomes: forest,

savanna, grassland, and desert. Their location and extent are influenced by both climate and human activities. Vegetation and soil, in turn, influence the types of agriculture that people practice in a particular region.

Physical Processes: Soil. Soil, the material that forms on Earth's surface, is the thin interface between the air and the rocks. Not merely dirt, soil contains the nutrients necessary for successful growth of plants, including those useful to humans.

The U.S. Comprehensive Soil Classification System divides global soil types into ten orders. The orders are subdivided into suborders, great groups, subgroups, families, and series. More than 12,000 soil types have been identified in the United States alone. Two basic problems contribute to the destruction of soil: erosion and depletion of nutrients.

(28)

Physical Processes: Landforms. Geographers find that the study of Earth's landforms—a science known as geomorphology—helps to explain the distribution of people and the choice of economic activities at different locations.

Geographers use topographic maps to study the relief and slope of localities. Relief is the difference in elevation between any two points, and it measures the extent to which an area is flat or hilly.

The Netherlands: Sensitive Environmental Modification. Few regions have been as thoroughly modified by humans as the Netherlands. More than half of the country lies below sea level. The building of dikes and polders began in the thirteenth century as private enterprise and has been continued by the government during the last 200 years. A polder is created by draining water from an area of land. In the north, a dike built in 1932 has turned the Zuider Zee from a saltwater sea to a freshwater lake. An ambitious 30-year project in the southwest, begun after a devastating flood in 1953, built dams to close off most of the water ways in the huge Delta formed by the Rhine, the Maas, and the Scheldt rivers. With these two massive projects finished, attitudes toward modifying the environment have changed in the Netherlands. The Dutch are deliberately breaking some dikes to flood fields. But modifying the environment will still be essential to the survival of the Dutch.

(29)

Florida: Not-So-Sensitive Environmental Modification. The fragile landscape of south Florida has been altered in insensitive ways, especially the barrier islands, the Everglades wetlands, and the Kissimmee River. The barrier islands are essentially large sandbars that shield the mainland from flooding and storm damage.

Despite their fragile condition, the barrier islands are attractive locations for constructing homes and recreational facilities. People build seawalls and jetties to fight erosion, but these projects result in more damage than protection. A seawall causes erosion on the down-current side of the island, by trapping sand along the up-current side. During the late 1940s the Army Corp of Engineers drained the northern third of the Everglades, opening 750,000 acres of land for growing sugarcane. The southern 1.4 million acres became a National Park. To protect the sugarcane fields and southern Florida cities from flooding, the Corps also built an elaborate set of levees, canals, and pumping stations. As a result, most of the freshwater that once reached the southern Everglades was pumped out to sea, and what water did reach the National Park was high in phosphorus, threatening native vegetation and endangering rare birds and other animals. A 1999 plan called for removing 60,000 acres from sugarcane production and pumping fresh water into the Park, rather than out to sea, but the survival of plants and animals of the Everglades now still depends on sensitive human management of the region's water flow.

The state of Florida asked the Army Corps of Engineers to straighten the course of the Kissimee River into a canal, because summer flooding rains were an obstacle to cattle grazing and urban growth. After the opening of the canal, polluted water mainly from cattle grazing along the banks ran into the canal and flowed into Lake Okeechobee, which is the source of freshwater for half of Florida's population. The Corps is now spending hundreds of millions of dollars to restore the Kissimmee River to its meandering course and to buy the nearby grazing land, which will again be subject to flooding.

Key Issue 3. Why Are Different Places Similar?
- **Scale: From local to global**
- **Space: Distribution of features**
- **Connections between places**

Although accepting that each place or region on Earth is unique, geographers recognize that human activities are rarely confined to one location. This section discusses three basic concepts—scale, space, and connections—that help geographers understand why two places or regions can display similar features.

(31)
Scale: From Local to Global
All scales from local to global are important in geography—the appropriate scale depends on the specific subject. At a local scale, such as a neighborhood within a city, geographers tend to see unique features. At the global scale, encompassing the entire world, geographers tend to see broad patterns. Geography matters in the contemporary world because it can explain human actions at all scales, from local to global.

Globalization of Economy
Scale is an increasingly important concept in geography because of **globalization**, which is a force or process that involves the entire world and results in making something worldwide in scope. Globalization means that the scale of the world is shrinking—not literally in size, of course, but in the ability of a person, object, or idea to interact with a person, object, or idea in another place.

A few people living in very remote regions of the world may be able to provide all of their daily necessities. But most economic activities undertaken in one region are influenced by interaction with decision makers located elsewhere. Globalization of the economy has been led primarily by **transnational corporations**, sometimes called multinational corporations. Modern technology provides the means to easily move money—as well as materials, products, technology, and other economic assets—around the world. Every place in the world is part of the global economy, but globalization has led to more specialization at the local level.

A locality may be especially suitable for a transnational corporation to conduct research, to develop new engineering systems, to extract raw material, to produce parts, to store finished products, to sell them, or to manage operations. Globalization of the economy has heightened economic differences among places.

(32)
Globalization of Culture
Geographers observe that increasingly uniform cultural preferences produce uniform "global" landscapes of material artifacts and of cultural values. The survival of a local culture's distinctive beliefs, forms, and traits is threatened by interaction with such social customs as wearing jeans and Nike shoes, consuming Coca-Cola and McDonalds hamburgers, and other preferences in food, clothing shelter, and leisure activities. Yet despite globalization, cultural differences among places not only persist but actually flourish in many places. The communications revolution that promotes globalization of culture also permits preservation of cultural diversity.

With the globalization of communications, people in two distant places can watch the same television program. At the same time, with the fragmentation of the broadcasting market, two people in the same house can watch different programs. Culturally, people residing in different places are displaying fewer differences and more similarities in their cultural preferences. But the desire of some people to retain their traditional cultural elements has led to political conflict and market fragmentation in some regions.

Strong determination on the part of a group to retain its local cultural traditions in the face of globalization of culture can lead to intolerance of people who display other beliefs, social forms, and material traits. Human geographers understand that many contemporary social problems result from a tension between forces promoting global culture and economy on the one hand and, on the other, preservation of local economic autonomy and cultural traditions.

(33)
Space: Distribution Features
Geographers think about the arrangements of people and activities found in space and try to understand why those people and activities are distributed across space as they are.

Distribution
The arrangement of a feature in space is known as distribution. Geographers identify three main properties of distribution across Earth: density, concentration, and pattern.

Density. The frequency with which something occurs in space is its density. **Arithmetic density**, which is the total number of objects in an area, is commonly used to compare the distribution of population in different countries. Arithmetic density involves two measures: the number of people and the land area. A large *population* does not necessarily lead to a high *density*. **Physiological density** is the number of persons per unit of area suitable for agriculture. The number of farmers per unit of area is called **agricultural density**. **Housing density** is the number of dwelling units per unit of area.

Concentration. The extent of a feature's spread over space is the concentration. *Clustered* objects in an area are close together, *dispersed* are far apart. Geographers use concentration to describe changes in distribution.

(35)

Pattern. Some features are organized in a geometric pattern, while others are distributed irregularly.

Gender and Ethnic Diversity in Space
Spatial interaction may be limited even among people in close proximity to one another. Consider first the daily movement of an "all-American" family.

Dad drives to work, spends the day and drives home. The mother's local-scale travel patterns are likely to be far more complex than the father's, as she transports children and organizes family life. Most American women are now employed at work outside the home, adding a substantial complication to an already complex pattern of moving across urban space. If the hypothetical family consisted of persons of color, its connections with space would change. In most U.S. neighborhoods the residents are virtually all whites or virtually all persons of color. Segregation persists partly based on cultural preference, partly based on fears.

Cultural identity is a source of pride to people at the local scale and an inspiration for personal values. Even more than self-identification, personal traits matter to other people. For geographers, concern for cultural diversity is not merely a politically correct expediency; it lies at the heart of geography's spatial tradition.

(36)
Connections between Places
Geographers apply the term space-time compression to describe the reduction in the time it takes for something to reach another place. Geographers explain the process, called diffusion, by which connections are made between regions, as well as the mechanism by which connections are maintained through networks.

Spatial Interaction
In the past, most forms of interaction among cultural groups required the physical movement of settlers, explorers, and plunderers from one location to another. Today travel by motor vehicle or airplane is much quicker and we can communicate instantly with people in distant places. When places are connected to each other through a network, geographers say there is spatial interaction between them.

Networks are chains of communication that connect places. A well-known example is the television network. Transportation systems also form networks that connect places to each other. Typically, the farther away one group is from another, the less likely the two groups are to interact. This trailing-off phenomenon is called distance decay.

(38)
Diffusion
Diffusion is the process by which a characteristic spreads across space from one place to another over time. The place from which an innovation originates is called a hearth. The dominant cultural, political, and economic features of contemporary United States and Canada can be traced primarily to hearths in Europe and the Middle East. Other regions of the world also contain important hearths. An idea, such as agriculture, may originate independently in more than one hearth. Geographers observe two basic types of diffusion: relocation and expansion.

Relocation Diffusion. The spread of an idea through physical movement of people is termed relocation diffusion. Relocation diffusion can explain the rapid rise in the number of AIDS cases in the United State during the 1980s and early 1990s but not the rapid decline beginning in the mid-1990s. The decline resulted from the rapid diffusion of preventive methods and medicines. The rapid spread of these innovations is an example of expansion diffusion rather than relocation diffusion.

Expansion Diffusion. The spread of a feature from one place to another in a snowballing process is **expansion diffusion**. This expansion may result from one of three processes: hierarchical diffusion, contagious diffusion, (and) stimulus diffusion. **Hierarchical diffusion** is the spread of an idea from persons or nodes of authority or power to other persons or places. **Contagious diffusion** is the rapid, widespread diffusion of a characteristic throughout the population. **Stimulus diffusion** is the spread of an underlying principle, even though a characteristic itself apparently fails to diffuse. Modern methods of communications . . . encourage . . . hierarchical diffusion. The Internet . . . has encouraged . . . contagious diffusion. All the new technologies support . . . stimulus diffusion.

(40)
Diffusion of Culture and Economy. In a global culture and economy, transportation and communications systems have been organized to rapidly diffuse raw materials, goods, services, and capital from nodes of origin to other regions. The global culture and economy is increasingly centered on three core or hearth regions of North America, Western Europe, and Japan. The global economy has produced greater disparities than in the past between the levels of wealth and well-being enjoyed by people in the core and in the periphery. The increasing gap in economic conditions . . . is known as **uneven development**.

Key Terms

Agricultural density (p.34)
Arithmetic density (p.34)
Base line (p.12)
Cartography (p.6)
Concentration (p.34)
Connections (p.5)
Contagious diffusion (p.40)
Cultural ecology (p.25)
Cultural landscape (p.20)
Culture (p.24)
Density (p.34)
Diffusion (p.38)
Distance decay (p.37)
Distribution (p.34)
Environmental determinism (p.25)
Expansion diffusion (p.38)
Formal region (p.20)
Functional region (p.22)
Geographic Information System (GIS) (p.14)
Global Positioning System (GPS) (p.14)
Globalization (p.31)
Greenwich Mean Time (p.19)

Hearth (p.38)
Hierarchical diffusion (p.39)
International Date Line (p.20)
Land Ordinance of 1785 (p.12)
Latitude (p.18)
Location (p.15)
Longitude (p.18)
Map (p.5)
Mental map (p.22)
Meridian (p.18)
Parallel (p.18)
Pattern (p.35)
Physiological density (p.34)
Place (p.5)
Polder (p.28)
Possibilism (p.26)
Prime meridian (p.18)
Principal meridian (p.12)
Projection (p.12)
Region (p.5)
Regional studies (p.20)
Relocation diffusion (p.38)
Remote sensing (p.14)

Chapter 2. Population

A study of population is the basis for understanding a wide variety of issues in human geography. Therefore, our study of human geography begins with a study of population.

Key Issues
1. Where is the world's population distributed?
2. Where has the world's population increased?
3. Why is population increasing at different rates in different countries?
4. Why might the world face an overpopulation problem?

(47)

The study of population is critically important for three reasons:
- More people are alive at this time—more than 6 billion—than at any time in human history;
- The world's population increased at a faster rate during the second half of the twentieth century than ever before in history;
- Virtually all global population growth is concentrated in less developed countries.

The scientific study of population characteristics is **demography**. At a global *scale*, . . . the world's so-called **overpopulation** problem is not simply a matter of the total number of people . . . but the relationship between number of people and available resources. At a local scale, geographers find that overpopulation is a threat in some regions of the world but not in others. Regions with the most people are not necessarily the same as the regions with an unfavorable balance between population and resources.

Key Issue 1. Where Is the World's Population Distributed?
- **Population concentrations**
- **Sparsely populated regions**
- **Population density**

We can understand how population is distributed by examining two basic properties: concentration and density.

Population Concentrations
Approximately two-thirds of the world's population is clustered in four regions: East Asia, South Asia, Southeast Asia, and Western Europe. The four regions display some similarities. Most of their people live near an ocean or near a river with easy access to an ocean. The four population clusters occupy generally low-lying areas, with fertile soil and temperate climate. Despite these similarities, we can see significant differences in the pattern of occupancy of the land.

(49)
East Asia
One-fifth of the world's people live in East Asia, the largest cluster of inhabitants. Five-sixths of the people in this concentration live in the People's Republic of China, the world's most populous country. The Chinese population is clustered near the Pacific Coast and in several fertile river valleys. Three fourths of the people live in rural areas where they work as farmers.

In Japan and South Korea, population is not distributed uniformly either. More than three-fourths of the Japanese and Koreans live in urban areas.

South Asia

The second-largest concentration of people, more than one-fifth, is in South Asia. India, the world's second most populous country, contains more than three-fourths of the South Asia population concentration. Much of this area's population is concentrated along the plains of the Indus and Ganges rivers. Population is also concentrated near India's two long coastlines. Like the Chinese, most people in South Asia are farmers.

Southeast Asia

The world's fourth-largest population cluster, after Europe, is in Southeast Asia, mostly on a series of islands. Indonesia, which consists of 13,677 islands, is the world's fourth most populous country. Several Philippine islands contain high population concentrations. The Indochina population is clustered along several river valleys and deltas at the southeastern tip of the Asian mainland. A high percentage of people in Southeast Asia work as farmers.

The three Asian population concentrations together comprise over half of the world's total population, but together they live on less than 10 percent of Earth's land area. The same held true 2,000 years ago.

Europe

Combining the populations of Western Europe, Eastern Europe, and the European portion of Russia forms the world's third-largest population cluster, one-ninth of the world's people. Three-fourths of Europe's inhabitants live in cities. They import food and other resources. The search for additional resources was a major incentive for Europeans to colonize other parts of the world during the previous six centuries.

Other Population Clusters

The largest population concentration in the Western Hemisphere is in the northeastern United States and southeastern Canada. About 2 percent of the world's people live in the area. Less than 5 percent are farmers.

Another 2 percent of the world's population is clustered in West Africa, especially along the south-facing Atlantic coast. Approximately half is in Nigeria, and the other half is divided among several small countries west of Nigeria. Most people work in agriculture.

(50)
Sparsely Populated Regions

Relatively few people live in regions that are too dry, too wet, too cold, or too mountainous for . . . agriculture. Approximately three-fourths of the world's population lives on only 5 percent of Earth's surface. The portion of the Earth's surface occupied by permanent human settlement is called the ecumene. The areas of Earth that humans consider too harsh for occupancy have diminished over time, while the ecumene has increased. Even 500 years ago much of North America and Asia lay outside the ecumene.

Dry Lands

Areas too dry for farming cover approximately 20 percent of Earth's land surface. Deserts generally lack sufficient water to grow crops . . . although some people survive there by raising animals, such as camels, that are adapted to the climate. Dry lands . . . may contain natural resources . . . notably, much of the world's oil reserves.

Wet Lands

Lands that receive very high levels of precipitation may also be inhospitable for human occupation. These lands are located primarily near the equator. The combination of rain and heat rapidly depletes nutrients from the soil, thus hindering agriculture. In seasonally wet lands, such as those in Southeast Asia, enough food can be grown to support a large population.

(51)
Cold Lands

Much of the land near the North and South poles is perpetually covered with ice or the ground is permanently frozen (permafrost). Few animals can survive the extreme cold, and few humans live there.

High Lands
Relatively few people live at high elevations. We can find some significant exceptions, especially in Latin America and Africa.

Population Density
Density . . . the number of people occupying an area of land, can be computed in several ways, including arithmetic density, physiological density, and agricultural density.

Arithmetic Density. Geographers most frequently use **arithmetic density**, which is the total number of people divided by total land area. (This measure is also called *population density*.) Arithmetic density enables geographers to make approximate comparisons of the number of people trying to live on a given piece of land in different regions of the world.

(52)
Physiological Density. A more meaningful population measure is afforded by looking at the number of people per area of a certain type of land in a region. Land suited for agriculture is called *arable land*. The number of people supported by a unit area of arable land is called the physiological density. Comparing physiological and arithmetic densities helps geographers to understand the capacity of the land to yield enough food for the needs of people.

Agricultural Density. Two countries can have similar physiological densities, but they may produce significantly different amounts of food because of different economic conditions. **Agricultural density** is the ratio of the number of farmers to the amount of arable land. To understand the relationship between population and resources in a country, geographers examine its physiological and agricultural densities together. The Netherlands has a much higher physiological density than does India but a much lower agricultural density.

(53)
Key Issue 2. Where Has the World's Population Increased?
- **Natural increase**
- **Fertility**
- **Mortality**

Population increases rapidly in places where many more people are born than die, increases slowly in places where the number of births exceeds the number of deaths by only a small margin, and declines in places where deaths outnumber births.

The population of a place also increases when people move in and decreases when people move out. This element of population change—migration—is discussed in Chapter 3. Geographers . . . measure population change . . . through three measures:
Crude birth rate (CBR) is the total number of live births in a year for every 1,000 people; **Crude death rate (CDR)** is the total number of deaths in a year for every 1,000 people; (and) **Natural increase rate (NIR)** is the percentage by which a population grows in a year.

The term *natural* means that a country's growth rate excludes migration. The term *crude* means that we are concerned with society as a whole rather than a refined look at particular individuals or groups.

Natural Increase
The world natural increase rate during the first decade of the twenty-first century is 1.3 percent. The world NIR is lower today than at its all-time peak of 2.2 percent in 1963. However, the NIR during the second half of the twentieth century was high by historical standards. The number of people added each year has dropped much more slowly than the NIR, because the population base is much higher

now than in the past. The rate of natural increase affects the **doubling time**, which is the number of years needed to double a population. When the NIR was 2.2 percent back in 1963, doubling time was 35 years.

(54)
Virtually 100 percent of the natural increase is clustered in less developed countries. To explain these differences in growth rates, geographers point to the regional differences in fertility and mortality rates.

Fertility
The highest crude birth rates are in sub-Saharan Africa, and the lowest are in Europe. Geographers also use the **total fertility rate (TFR)** to measure the number of births in a society. The TFR is the average number of children a woman will have throughout her childbearing years. The total fertility rate for the world as a whole is approximately three. The TFR exceeds six in many countries of sub-Saharan Africa, compared to less than two in nearly every European country.

Mortality
Two useful measures of mortality in addition to the crude death rate . . . are the infant mortality rate and life expectancy. The **infant mortality rate (IMR)** is the annual number of deaths of infants under one year of age, compared with total live births . . . usually expressed as the number of deaths . . . per 1,000 births rather than as a percentage. Infant mortality rates exceed 100 in some LDCs. In general, the IMR reflects a country's health-care system.

(56)
Minorities in the United States have infant mortality rates that are twice as high as the national average, comparable to levels in Latin America and Asia.

Life expectancy at birth measures the average number of years a newborn infant can expect to live at current mortality levels. Babies born today can expect to live into their late seventies in Western Europe but only into their late forties in many sub-Saharan African countries.

Higher rates of natural increase, crude birth, total fertility, and infant mortality, and lower life expectancy are in LDCs. The final world map of demographic variables—crude death rate—does not follow the familiar pattern. The combined crude death rate for all less developed countries is actually lower than the combined rate for MDCs. The populations of different countries are at various stages in an important process known as the demographic transition.

(58)
Key Issue 3. Why Is Population Increasing at Different Rates in Different Countries?
- **The demographic transition**
- **Population pyramids**
- **Countries in different stages of demographic transition**
- **Demographic transition and world population growth**

While (population) rates vary among countries, a similar process of change . . . is operating, known as the **demographic transition**.

The Demographic Transition
The demographic transition (has four stages) . . . and—barring a catastrophe such as a nuclear war—it is irreversible.

Stage 1: Low Growth
Most of humanity's several-hundred-thousand-year occupancy of Earth was characterized by stage 1 of the demographic transition. Crude birth and death rates varied considerably from one year to the next and from one region to another, but over the long term they were roughly comparable, at very high levels.

(59)

Between 8000 B.C. and A.D. 1750, Earth's human population increased from approximately 5 million to 800 million. The burst of population growth around 8000 B.C. was caused by the **agricultural revolution**. Despite the agricultural revolution, the human population remained in stage 1 of the demographic transition because food supplies were still unpredictable.

Stage 2: High Growth

For nearly 10,000 years after the agricultural revolution, world population grew at a modest pace. After around A.D. 1750 the world's population suddenly began to grow 10 times faster than in the past. In stage 2, the crude death rate suddenly plummets, while the crude birth rate remains roughly the same as in stage 1.

Some demographers divide stage 2 into two parts. During the second part, the growth rate begins to slow, although the gap between births and deaths remains high. Countries entered stage 2 of the demographic transition after 1750 as a result of the **Industrial Revolution**. . . . The result of this transformation was an unprecedented level of wealth, some of which was used to make communities healthier places to live. Countries in Europe and North America entered stage 2 of the demographic transition about 1800, but stage 2 did not diffuse to most countries in Africa, Asia, and Latin America until around 1950.

(60)

The late twentieth-century push of countries into stage 2 was caused by the **medical revolution**. Improved medical practices suddenly eliminated many of the traditional causes of death in LDCs and enabled more people to experience longer and healthier lives.

Stage 3: Moderate Growth

A country moves from stage 2 into stage 3 of the demographic transition when the crude birth rate begins to drop sharply. European and North American countries . . . moved from stage 2 to stage 3 . . . during the first half of the twentieth century. Most countries in Asia and Latin America have moved to stage 3 in recent years, while most of African countries remain in stage 2. A society enters stage 3 . . . when people choose to have fewer children. Medical practices introduced in stage 2 societies greatly improved the probability of infant survival, but many years elapsed before families reacted by conceiving fewer babies. Economic changes in stage 3 societies also induce people to have fewer offspring. Farmers often consider a large family to be an asset. In contrast, children living in cities are generally not economic assets.

Stage 4: Low Growth

A country reaches stage 4 . . . when the crude birth rate declines to the point where it equals the crude death rate. The condition is called **zero population growth (ZPG)**. Demographers more precisely define zero population growth as the total fertility rate (TFR) that results in a lack of change in the total population over a long term. A TFR of approximately 2.1 produces ZPG, although a country that receives many immigrants may need a lower total fertility rate to achieve ZPG.

Most European countries have reached stage 4 of the demographic transition. The United States has moved slightly below ZPG since 2000. When most families lived on farms, employment and child rearing were conducted at the same place, but in urban societies most parents must leave the home to work. Changes in lifestyle also encourage smaller families. Several Eastern European countries, most notably Russia, have negative natural increase rates, . . . a legacy of a half century of Communist rule.

(61)

As memories of the Communist era fade, Russians and other Eastern Europeans may display birth and death rates more comparable to those in Western Europe. Alternatively, demographers in the future may identify a fifth stage, . . . characterized by higher death rates than birth rages and an irreversible population decline.

The Demographic Transition in England

England has reached stage 4, and at least fragmentary information on its population is available for the past 1,000 years.

Stage 1: Low Growth until 1750. In 1066, when the Normans invaded England, the country's population was approximately 1 million. Seven hundred years later the population was only 6 million. During that 700-year period, population rose in some years and fell in others. As recently as the 1740s, the crude death rate skyrocketed following a series of bad harvests.

Stage 2: High Growth (1750–1880). In 1750 the crude birth and death rates in England were both 40 per 1,000. In 1800 the crude birth rate remained very high at 34, but the crude death rate had plummeted to 20. England remained in stage 2 . . . for about 125 years.

Stage 3: Moderate Growth (1880–early 1970s). The crude death rate continued to fall somewhat over the next century, from 19 per 1,000 in 1880 to 12 in 1970. However, the crude birth rate declined sharply, from 33 per 1,000 in 1880 to 18 by 1930 and 15 in 1970.

(62)

Stage 4: Low Growth (Early 1970s–present). Since the early 1970s . . . total fertility rate has long been below the 2.1 needed for replacement. England's population has grown by 3 million since 1970, primarily because of immigration from former colonies.

Population Pyramids

Population in a country is influenced by the demographic transition in two principal ways: the percentage of the population in each age group, and the distribution of males and females. A country's population can be displayed by age and gender groups on a bar graph called a **population pyramid**.

Age Distribution

The age structure of a population is extremely important in understanding similarities and differences among countries. The most important factor is the **dependency ratio**, which is the number of people who are too young or too old to work, compared to the number of people in their productive years. Young dependents outnumber elderly ones by 10:1 in stage 2 countries, but the numbers of young and elderly dependants are roughly equal in stage 4 countries. The large percentage of children in sub-Saharan Africa and other stage 2 countries strains the ability of poorer countries to provide needed services. As countries pass through the stages of the demographic transition, the percentage of elderly people increases. More than one-fourth of all government expenditures in the United States, Canada, Japan, and many European countries go to Social Security, health care, and other programs for the older population.

Sex Ratio

The number of males per hundred females in the population is the sex ratio. In Europe and North America the ratio of men to women is about 95:100. In the rest of the world the ratio is 102:100.

(63)

In poorer countries the high mortality rate during childbirth partly explains the lower percentage of women. The difference also relates to the age structure. Societies with a high rate of immigration typically have more males than females. Retirement communities have relatively high percentages of women, because they have longer life expectancies. The shape of a community's population pyramid tells a lot about its distinctive character. The different shapes result from differences in the ethnic composition.

Countries in Different Stages of Demographic Transition

No country today remains in stage 1 of the demographic transition, but it is instructive to compare countries in each of the other three stages.

Cape Verde: Stage 2 (High Growth). Cape Verde, a collection of 12 small islands in the Atlantic Ocean off the coast of West Africa, moved from stage 1 to stage 2 about 1950.

(64)

During the first half of the twentieth century Cape Verde's population declined. The large gap between births and deaths most years produced a high natural increase rate typical of stage 2, yet Cape Verde remained in stage 1 until 1950. Famines dramatically disrupted the typical patterns of birth, death, and natural increase. During the half century since entering stage 2, the population of Cape Verde has nearly tripled, to approximately 400,000.

Cape Verde moved on to stage 2 when an antimalarial campaign was launched.

(66)

Meanwhile, as is typical of stage 2 countries, Cape Verde's crude birth rate has remained relatively high and still fluctuates wildly. The wild fluctuations in Cape Verde's crude birth rate are a legacy of the severe famine during the 1940s. The population pyramid shows that Cape Verde has a large number of females age 5–14 who will soon start moving into their prime childbearing years. For Cape Verde to enter stage 3 . . . these females must bear considerably fewer children than did their mothers.

Chile: Stage 3 (Moderate Growth). Like most countries outside Europe and North America, Chile entered the twentieth century still in stage 1. Much of Chile's population growth—as in other countries in the Western Hemisphere—resulted from European immigration. Chile's crude death rate declined sharply in the 1930s, moving the country into stage 2. Chile's crude death rate was lowered by the infusion of medical technology from MDCs. Chile has been in stage 3 since about 1960 . . . primarily because of a vigorous government family planning policy.

(67)

Reduced income and high unemployment . . . also induced couples to . . . delay childbearing. The country is unlikely to move into stage 4 in the near future. Chile's government reversed its policy and renounced support for family planning during the 1970s. The government policy was that population growth could help promote national security and economic development. Also . . . most Chileans belong to the Roman Catholic Church, which opposes the use of what it calls artificial birth-control techniques.

Denmark: Stage 4 (Low Growth). Denmark's history is similar to that of England's. The country entered stage 2 . . . in the nineteenth century, when the crude death rate began its permanent decline. The crude birth rate then dropped in the late nineteenth century, and the country moved on to stage 3.

(68)

Since the 1970s the country has reached zero population growth, and the population is increasing almost entirely because of immigration. Instead of a classic pyramid shape, Denmark has a column, demonstrating that the percentages of young and elderly people are nearly the same.

Demographic Transition and World Population Growth

Worldwide population increased rapidly during the second half of the twentieth century. The four-stage demographic transition is characterized by two big breaks with the past. The first break—the sudden drop in the death rate—has been accomplished everywhere. The second break—the sudden drop in the birth rate—has yet to be achieved in many countries. The nineteenth-century decline in the CDR in Europe and North America took place in conjunction with the Industrial Revolution. In contrast, the sudden drop in the CDR in Africa, Asia, and Latin America in the twentieth century was accomplished by different means and with less internal effort by local citizens.

(69)

Medical technology was injected from Europe and North America instead of arising within the country as part of an economic revolution. In the past, stage 2 lasted for approximately 100 years in Europe and North America, but today's stage 2 countries are being asked to move through to stage 3 in much less time in order to curtail population growth.

Key Issue 4. Why Might the World Face an Overpopulation Problem?
- **Malthus on overpopulation**
- **Declining birth rates**
- **World Health Threats**

Why does global population growth matter? Geographers observe that diverse local culture and environmental conditions may produce different answers in different places.

Malthus on Overpopulation
English economist Thomas Malthus (1766–1834) was one of the first to argue that the world's rate of population increase was far outrunning the development of food supplies.

Population Growth vs. Food Supply
In *An Essay on the Principle of Population*, published in 1798, Malthus claimed that population . . . increased geometrically, while food supply increased arithmetically.

He concluded that population growth would press against available resources in every country, unless "moral restraint" produced lower crude birth rates or unless disease, famine, war, or other disasters produced higher crude death rates.

Neo-Malthusians. Contemporary geographers and other analysts are taking another look at Malthus's theory, because of the unprecedented rate of natural increase in LDCs. Neo-Malthusians paint a frightening picture of a world in which billions of people are engaged in a desperate search for food and fuel. Many LDCs have expanded their food production significantly in recent years, but they have more poor people than ever before.

(70)

Malthus's Critics. Criticism has been leveled at both the population growth and resource depletion sides of Malthus's equation. Contemporary analysts such as Esther Boserup and Julian Simon (argue that) a larger population could stimulate economic growth and therefore the production of more food. The Marxist theorist Friedrich Engels dismissed Malthus's arithmetic as an artifact of capitalism. Engels argued that the world possessed sufficient resources to eliminate global hunger and poverty, if only these resources were shared equally.

Malthus' critics argue that the world is better off economically with 6 plus billion people than it was with 1 billion, because too few people can retard economic development. Some political leaders, especially in Africa, argue that more people will result in greater power.

Declining Birth Rates
Although the Malthus theory seems unduly pessimistic on a global scale, geographers recognize the diversity of conditions among regions of the world. Although the world as a whole may not be in danger of "running out" of food, some regions with rapid population growth do face shortages of food.

Malthus Theory and Reality
Vaclav Smil has shown that Malthus was fairly close to the mark on food production but much too pessimistic on population growth.

(71)

Many people in the world cannot afford to buy food or do not have access to sources of food, but these are problems of distribution of wealth rather than insufficient global production of food, as Malthus theorized.

Population has been increasing at a much slower rate during the past two decades than it was during the previous half-century. However, neo-Malthusians point out that despite the lower NIR during the 1990s, the world added approximately the same number of people as during the 1980s.

Reasons for Declining Birth Rates

The natural increase can decline for only two reasons: lower birth rates or higher death rates. In most countries, the decline has occurred because of a lower birth rate, but in some countries of sub-Saharan Africa, the crude death rate is increasing.

Two strategies have been successful in reducing birth rates. One alternative emphasizes reliance on economic development, the other on distribution of contraceptives. Because of varied economic and cultural conditions, the most effective method varies among countries.

Economic Development. One approach . . . emphasizes . . . improving local economic conditions. If more women are able to attend school, . . . they . . . learn employment skills, gain more economic control of their lives, and make more informed reproductive choices. With the survival of more infants assured, women would be more likely to choose . . . contraceptives to limit the number of children.

(72)

Distribution of Contraceptives. In less developed countries, demand for contraceptive devices is greater than the available supply. About one-fourth of African women employ contraceptives, compared to about two-thirds in other less developed countries. Very high birth rates in Africa and southwestern Asia also reflect the relatively low status of women.

Many oppose birth-control programs for religious and political reasons. Analysts agree that the most effective means of reducing births would employ both alternatives. But LDC governments and international family-planning organizations have limited funds . . . so they must set priorities.

World Health Threats

Lower crude birth rates have been responsible for declining natural increase rates in most countries. However, in some countries of sub-Saharan Africa lower natural increase rates have also resulted from higher crude death rates, especially through the diffusion of AIDS. Medical researchers have identified an epidemiologic transition that focuses on distinctive causes of death in each stage of the demographic transition.

Epidemiologic Transition Stages 1 and 2

Stage 1 of the epidemiologic transition, as originally formulated by epidemiologist Abdel Omran in 1971, has been called the stage of pestilence and famine. Infectious and parasitic diseases were principal causes of human deaths.

Black Plague. The Black Plague, or bubonic plague, originated in present-day Kyrgyzstan and was brought from there by a Tatar army when it attacked an Italian trading post on the Black Sea. About 25 million Europeans died between 1347 and 1350, at least one-half of the continent's population. Five other epidemics in the late fourteenth century added to the toll in Europe. In China, 13 million died from the plague in 1380.

Cholera. Stage 2 of the epidemiologic transition has been called the stage of receding pandemics. A pandemic is disease that occurs over a wide geographic area and affects a very high proportion of the population.

(73)

Cholera became an especially virulent epidemic in urban areas during the Industrial Revolution. Dr. John Snow (1813-58) mapped the distribution of deaths from cholera in 1854 in the poor London neighborhood of Soho. He overlaid a map of the distribution of cholera victims with a map of the distribution of water pumps, and found that a large percentage of cholera victims were clustered around one pump, on Broad Street.

(74)

Construction of water and sewer systems eradicated cholera by the late nineteenth century. However, cholera reappeared a century later in rapidly growing cities of less developed countries as they moved into stage 2 of the demographic transition.

Epidemiologic Transition Stages 3 and 4

Stage 3 of the epidemiologic transition, the stage of degenerative and human-created diseases, is characterized by a decrease in deaths from infectious diseases and an increase in chronic disorders associated with aging. The two especially important chronic disorders in stage 3 are cardiovascular diseases, such as heart attacks, and various forms of cancer.

The decline in infectious diseases has been sharp in stage 3 countries. Effective vaccines were responsible for these declines. As less developed countries have moved recently from stage 2 to stage 3, infectious diseases have also declined.

Omran's epidemiologic transition was extended by S.Jay Olshansky and Brian Ault to stage 4, the stage of delayed degenerative diseases. The major degenerative causes of death—cardiovascular diseases and cancers—linger, but the life expectancy of older people is extended through medical advances.

Epidemiologic Transition Possible Stage 5

Some medical analysts argue that the world is moving into stage 5 of the epidemiologic transition, the stage of reemergence of infectious and parasitic diseases. Infectious diseases thought to have been eradicated or controlled have returned, and new ones have emerged.

Reasons for Stage 5. Three reasons help to explain the possible emergence of a stage 5 of the epidemiologic transition. One is evolution. Infectious disease microbes have continuously evolved and changed in response to environmental pressures by developing resistance to drugs. Malaria was nearly eradicated in the mid-twentieth century by spraying DDT in areas infested with the mosquito that carried the parasite. The disease returned after 1963, however, and now causes more than 2 million deaths worldwide. The reason was the evolution of DDT-resistant mosquitoes.

A second reason for continued epidemics is poverty. Tuberculosis (TB) is an example of an infectious disease that has been largely controlled in relatively developed countries like the United States but remains a major cause of death in less developed countries.

(75)
The third factor in the reemergence of epidemics is improved travel. As they travel, people carry diseases with them and are exposed to the diseases of others. An example is severe acute respiratory syndrome (SARS), which originated in China in November 2002 and diffused to other countries, such as Canada and Vietnam.

SARS graphically displayed both the risks and benefits of globalization. On the one hand, SARS diffused around the world within days. On the other hand, information about SARS also diffused rapidly around the world. Epidemiologists quickly traced the diffusion of the disease from Guangchou province China to a hotel in Hong Kong, then to other places in the world where guests of that hotel went.
(76)
Some fear that terrorists may also be responsible for spreading infectious diseases. After September 11, U.S. government officials urged health care and other emergency response workers to be immunized against smallpox, because terrorists were thought to have access to samples of the disease that remained for medical research.

AIDS. The most lethal epidemic in recent years has been AIDS (acquired immunodeficiency syndrome), caused by the human immunodeficiency virus (HIV). The impact of AIDS has been felt most strongly in sub-Saharan Africa. With about 11 percent of the world's population, sub-Saharan Africa has 70 percent of the world's HIV-positive population.

Other than South Africa, the country with the highest number HIV-positive in 2001 was India. The second highest rate of infection was in Caribbean countries, such as Haiti. Crude death rates in many sub-Saharan Africa countries rose sharply during the 1990s as a result of AIDS, from the mid-teens to the low twenties. The populations of Botswana and South Africa are forecast to decline between now and 2050 as a result of AIDS.

Key Terms

Agricultural density (p.52)
Agricultural revolution (p.59)
Arithmetic density (p.51)
Census (p.65)
Crude birth rate (CBR) (p.53)
Crude death rate (CDR) (p.53)
Demographic transition (p.58)
Demography (p.47)
Dependency ratio (p.62)
Doubling time (p.54)
Epidemiologic transition (p.72)
Epidemiology (p.72)
Ecumene (p.50)

Industrial Revolution (p.59)
Infant mortality rate (IMR) (p.55)
Life expectancy (p.57)
Medical Revolution (p.60)
Natural increase rate (NIR) (p.53)
Overpopulation (p.47)
Pandemic (p.72)
Physiological density (p.52)
Population pyramid (p.62)
Sex ratio (p.62)
Total fertility rate (TFR) (p.55)
Zero population growth (ZPG) (p.60)

Chapter 3. Migration

Geographers document *from where* people migrate and *to where* they migrate. They also study reasons *why* people migrate. Most people migrate in search of three objectives: economic opportunity, cultural freedom, and environmental comfort.

Key Issues
1. Why do people migrate?
2. Where are migrants distributed?
3. Why do migrants face obstacles?
4. Why do people migrate within a country?

(85)

The subject of this chapter is a specific type of relocation diffusion called **migration**, which is a permanent move to a new location. **Emigration** is migration *from* a location; **immigration** is migration *to* a location. The difference between the number of immigrants and the number of emigrants is the **net migration**. Migration is a form of **mobility**, which is a more general term covering all types of movements from one place to another. Short-term, repetitive, or cyclical movements that recur on a regular basis, such as daily, monthly, or annually, are called **circulation**.

If people can participate in the globalization of culture and economy regardless of place of residence, why do they still migrate in large numbers? The answer is that *place* is still important to an individual cultural identity and economic prospects.

Key Issue 1. Why Do People Migrate?
- **Reasons for migrating**
- **Distance of migration**
- **Characteristics of migrants**

Geography has no comprehensive theory of migration, although a nineteenth-century outline of 11 migration "laws" written by E. G. Ravenstein is the basis for contemporary migration studies. Ravenstein's "laws" can be organized into three groups: (reasons, distance, and migrant characteristics).

Reasons for Migrating
Most people migrate for economic reasons. Cultural and environmental factors also induce migration, although not as frequently as economic factors. People decide to migrate because of push factors and pull factors. A **push factor** induces people to move out of their present location, whereas a **pull factor** induces people to move into a new location. Both push and pull factors typically play a role. We can identify three major kinds of push and pull factors: economic, cultural, and environmental.

Economic Push and Pull Factors
Most people migrate for economic reasons. The relative attractiveness of a region can shift with economic change.

(86)
Cultural Push and Pull Factors
Forced international migration has historically occurred for two main cultural reasons: slavery and political instability. In the twentieth century, forced international migration increased because of political instability resulting from cultural diversity. Forced migration of ethnicities is discussed in more detail in Chapter 7. Refugees are people who have been forced to migrate from their home country and cannot return for fear of persecution.

Political conditions can also operate as pull factors, especially the lure of freedom. With the election of democratic governments in Eastern Europe during the 1990s, Western Europe's political pull has disappeared as a migration factor. However, Western Europe pulls an increasing number of migrants from Eastern Europe for economic reasons.

(87)
Environmental Push and Pull Factors
People also migrate for environmental reasons, pulled toward physically attractive regions and pushed from hazardous ones. Attractive environments for migrants include mountains, seasides, and warm climates. Migrants are also pushed from their homes by adverse physical conditions. Water—either too much or too little—poses the most common environmental threat.

(88)
Intervening Obstacles
Where migrants go is not always their desired destination. They may be blocked by an **intervening obstacle**. In the past, intervening obstacles were primarily environmental . . . like mountains and deserts. Bodies of water long have been important intervening obstacles. However, today's migrant faces intervening obstacles created by local diversity in government and politics.

Distance of Migration
Ravenstein's theories made two main points about the distance that migrants travel to their home: Most migrants relocate a short distance and remain within the same country. Long-distance migrants to other countries head for major centers of economic activity.

Internal Migration
International migration is permanent movement from one country to another, whereas **internal migration** is permanent movement within the same country. International migrants are much less numerous than internal migrants. **Interregional migration** is movement from one region of a country to another, while **intraregional migration** is movement within one region.

International Migration
International migration is further divided into two types: forced and voluntary.

(89)
Geographer Wilber Zelinsky has identified a **migration transition**, which consists of changes in a society comparable to those in the demographic transition. A society in stage 1, . . . unlikely to migrate permanently . . . does have high daily or seasonal mobility in search of food. According to migration transition theory, societies in stages 3 and 4 are the destinations of the international migrants leaving the stage 2 countries in search of economic opportunities. Internal migration within countries in stages 3 and 4 of the demographic transition is intraregional, from cities to surrounding suburbs.

Characteristics of Migrants
Ravenstein noted distinctive gender and family-status patterns in his migration theories: Most long-distance migrants are male; . . . (and) . . . Most long-distance migrants are adult individuals rather than families with children.

Gender of Migrants
But since the 1990s the gender pattern has reversed, and women now constitute about 55 percent of U.S. immigration.

Family Status of Migrants
Ravenstein also believed that most long-distance migrants were young adults seeking work. For the most part, this pattern continues for the United States. With the increase in women migrating . . . more children are coming with their mother.

The origin of Mexican immigrants to the United States matches the expectations of the migration transition and distance-decay theories. The destination of choice within the United States is overwhelmingly states that border Mexico. But most immigrants originate not from Mexico's northern states but from interior states.

(90)
Because farm work is seasonal . . . the greatest number of Mexicans head north to the United States in the autumn and return home in the spring.

Key Issue 2. Where Are Migrants Distributed?
- **Global migration patterns**
- **U.S. immigration patterns**
- **Impact of immigration on the United States**

About 3 percent of the world's people are international migrants. The country with by far the largest number . . . is the United States.

Global Migration Patterns
At a global scale, Asia, Latin America, and Africa have net out-migration, whereas North America, Europe, and Oceania have net in-migration.

The population of the United States includes about 30 million individuals born in other countries. Although it contains the largest number of immigrants, the United States has a smaller percentage of immigrants than many other countries. The highest percentage of immigrants can be found in the Middle East, about one-half of the region's total population.

(91)
U.S. Immigration Patterns
About 70 million people have migrated to the United States since 1820, including the 30 million currently alive.

The United States has had three main eras in immigration. The three eras have drawn migrants from different regions.

Colonial Immigration from England and Africa
Immigration to the American colonies and the newly independent United States came from two sources: Europe and Africa. Most of the Africans were forced to migrate to the United States as slaves, whereas most Europeans were voluntary migrants—although harsh economic conditions and persecution in Europe blurred the distinction between forced and voluntary migration for many Europeans.

(92)
Ninteenth-century Immigration from Europe
In the 500 years since Christopher Columbus sailed from Spain to the Western Hemisphere, about 65 million Europeans have migrated to other continents. For 40 million of them the destination was the United States.

First Peak of European Immigration. From 1607 . . . until 1840, a steady stream of Europeans (totaling 2 million) migrated to the American colonies and after 1776 . . . the United States. Ninety percent of European immigrants . . . prior to 1840 came from Great Britain. During the 1840s and 1850s, the level of immigration . . . surged. More than 4 million people migrated, . . . more than twice as many as in the previous 250 years combined.

(94)
More than 90 percent of all U.S. immigrants during the 1840s and 1850s came from Northern and Western Europe, including two fifths from Ireland and another one third from Germany.

Second Peak of European Immigration. U.S. immigration declined somewhat during the 1860s as a result of the Civil War (1861–1865). A second peak was reached during the 1880s, more than a half-million people annually. More than three-fourths of the immigrants during the late 1880s came from Northern and Western Europe.

Third Peak of European Immigration. Economic problems in the United States discouraged immigration during the early 1890s, but by the end of the decade the level reached a third peak. The record year was 1907, with 1.3 million. During the third peak, most came from countries that previously had sent few people. One-fourth each came from Italy, Russia, and Austria-Hungary. The shift coincided with the diffusion of the Industrial Revolution . . . to Southern and Eastern Europe.

Recent Immigration from Less Developed Regions
Immigration to the United States dropped sharply in the 1930s and 1940s, during the Great Depression and World War II, . . . steadily increased during the 1950s, 1960s, and 1970s, then surged during the 1980s and 1990s to historically high levels.

Immigration from Asia. Asia was the leading source of immigrants between the late 1970s and the late 1980s until overtaken by Latin America. Asians also comprise more than 40 percent of Canadian immigrants, but compared to the U. S., Canada receives a much higher percentage of Europeans and a lower percentage of Latin Americans.

Immigration from Latin America. About 2 million Latin Americans migrated to the United States between 1820 and 1960, about 11 million between 1960 and 2000.

(95)
The unusually large number of immigrants from Mexico and other Latin American countries in 1990 and 1991 resulted from the 1986 Immigration Reform and Control Act, which issued visas to several hundred thousand who had entered the United States in previous years without legal documents.

Impact of Immigration on the United States
The U.S. population has been built up through a combination of emigration from Africa and England primarily during the eighteenth century, from Europe primarily during the nineteenth century, and from Latin America and Asia primarily during the twentieth century. In the twenty-first century, the impact of immigration varies around the country.

Legacy of European Migration
Massive European migration ended with the start of World War I.

(96)
Europe's Demographic Transition. Rapid population growth in Europe fueled emigration, especially after 1800. Application of new technologies . . . pushed much of Europe into stage 2 of the demographic transition. To promote more efficient agriculture, some European governments forced the consolidation of several small farms into larger units. Displaced farmers could choose between working in factories in the large cities or migrating to the United States or another country where farmland was plentiful.

Diffusion of European Culture. Europeans frequently imposed political domination on existing populations and injected their cultural values with little regard for local traditions. Economies in Africa and Asia became based on extracting resources for export to Europe, rather than on using those resources to build local industry. Many of today's conflicts in former European colonies result from past practices by European immigrants.

Undocumented Immigration to the United States
Many people who cannot legally enter the United States are now immigrating illegally, . . . called **undocumented immigrants**. The U.S. Bureau of Citizenship and Immigration Services (BCIS)

estimate 7 million undocumented immigrants in the U.S., although other estimates are as high as 20 million.

The BCIS apprehends more than a million persons annually trying to cross the southern U.S. border. Half of the undocumented residents legally enter the country as students or tourists and then remain after they are supposed to leave.

(97)
The 1986 Immigration Reform and Control Act tried to reduce the flow of illegal immigrants. Aliens who could prove that they had lived in the United States continuously between 1982 and 1987 could become permanent resident aliens and apply for U.S. citizenship after 5 years. At the same time, the law discouraged further illegal immigration by making it harder for recent immigrants to get jobs without proper documentation.

Destinations of Immigrants within the United States
Recent immigrants are not distributed uniformly through the United States. **Chain migration** is the migration of people to a specific location because relatives or members of the same nationality previously migrated there.

(98)
Key Issue 3. Why Do Migrants Face Obstacles?
- **Immigration policies of host countries**
- **Cultural problems living in other countries**

The major obstacles faced by most immigrants now begin only after they arrive . . . gaining permission to enter . . . and hostile attitudes of citizens.

(99)
Immigration Policies of Host Countries
The United States uses a quota system to limit the number . . . who can migrate. Other major recipients of immigrants . . . permit guest workers . . . but not to stay permanently.

U.S. Quota Laws
The era of unrestricted immigration to the United States, ended when Congress passed the Quota Act in 1921 and the National Origins Act in 1924. Quota laws were designed to assure that most immigrants to the United States continued to be Europeans. Quotas for individual countries were eliminated in 1968 and replaced with hemispheric quotas. In 1978 the hemisphere quotas were replaced by a global quota of 290,000, including a maximum of 20,000 per country. The current law has a global quota of 620,000, with no more than 7 percent from one country, but numerous qualifications and exceptions can alter the limit considerably.

Brain Drain. Other countries charge that by giving preference to skilled workers, U.S. immigration policy now contributes to a **brain drain**, which is a large-scale emigration by talented people. The average immigrant has received more education than the typical American: nearly one-fourth of all legal immigrants to the United States have attended graduate school, compared to less than one-tenth of native-born Americans.

Temporary Migration for Work
Prominent forms of temporary work migrants include **guest workers** in Europe and the Middle East and historically time-contract workers in Asia.

(100)
Foreign-born workers comprise more than one-half of the labor force in Luxembourg, one-sixth in Switzerland, and one-tenth in Austria, Belgium, and Germany.

Guest workers serve a useful role in Western Europe because they take low-status and low-skilled jobs that local residents won't accept. The United Kingdom severely restricts the ability of foreigners to obtain work permits. However, British policy is complicated by the legacy of the country's former worldwide empire. Most guest workers in Europe come from North Africa, the Middle East, Eastern Europe, and Asia.

Time-contract Workers
Millions of Asians migrated in the nineteenth century as time-contract laborers, recruited for a fixed period to work in mines or on plantations. More than 29 million ethnic Chinese currently live permanently in other countries, for the most part in Asia.

(101)
In recent years people have immigrated illegally in Asia to find work in other countries. Estimates of illegal foreign workers in Taiwan range from 20,000 to 70,000. Most are Filipinos, Thais, and Malaysians.

Distinguishing between Economic Migrants and Refugees
It is sometimes difficult to distinguish between migrants seeking economic opportunities and refugees fleeing from the persecution of an undemocratic government. The distinction between economic migrants and refugees is important, because the United States, Canada, and Western European countries treat the two groups differently.

Emigrants from Cuba. Since the 1959 revolution that brought the Communist government of Fidel Castro to power, the U.S. government has regarded emigrants from Cuba as political refugees. In the years immediately following the revolution, more than 600,000 Cubans were admitted to the United States. A second flood of Cuban emigrants reached the United States in 1980, when Fidel Castro suddenly decided to permit political prisoners, criminals, and mental patients to leave the country.

(102)
Emigrants from Haiti. Shortly after the 1980 Mariel boatlift from Cuba, several thousand Haitians also sailed in small vessels for the United States. Claiming that they had migrated for economic advancement, . . . U.S. immigration officials would not let the Haitian boat people stay. The Haitians brought a lawsuit. The government settled the case by agreeing to admit the Haitians. After a 1991 coup that replaced Haiti's elected president, Jean-Bertrand Aristide, thousands of Haitians fled their country . . . but the U.S. State Department decided that most left Haiti for economic rather than political reasons. The United States invaded Haiti in 1994 to reinstate Aristide as president. Many Haitians still try to migrate to the United States.

Emigrants from Vietnam. The long Vietnam War ended in 1975. The United States . . . evacuated from Saigon several thousand people. Thousands of other pro-U.S. South Vietnamese who were not politically prominent enough to get space on an American evacuation helicopter tried to leave by boat. A second surge of Vietnamese boat people began in the late 1980s. According to an international agreement, most of the Vietnamese boat people who were judged refugees were transferred to other places, especially the United States, Canada, Australia, and France. However, the majority of the boat people, who were considered economic migrants, were placed in detention camps . . . until 1996, when the camps were closed and the remaining boat people were sent back to Vietnam.

(103)
Vietnam remains a major source of immigrants to the United States, but the pull of economic opportunity in the United States is a greater incentive than the push of political persecution.

Cultural Problems Living in Other Countries
For many immigrants, admission to another country does not end their problems. Politicians exploit immigrants as scapegoats for local economic problems.

U.S. Attitudes toward Immigrants

Americans have always regarded new arrivals with suspicion but tempered their dislike during the nineteenth century because immigrants helped to settle the frontier and extend U.S. control across the continent. Opposition to immigration intensified when the majority of immigrants ceased to come from Northern and Western Europe. More recently, hostile citizens in California and other states have voted to deny undocumented immigrants access to most public services, such as schools, day-care centers, and health clinics.

Attitudes toward Guest Workers

In Europe, many guest workers suffer from poor social conditions. Both guest workers and their host countries regard the arrangement as temporary. In reality, however, many guest workers remain indefinitely, especially if they are joined by other family members. As a result of lower economic growth rates, Middle Eastern and Western European countries have reduced the number of guest workers in recent years. Political parties that support restrictions on immigration have gained support in France, Germany, and other European countries, and attacks by local citizens on immigrants have increased.

In the Middle East, petroleum-exporting countries fear that the increasing numbers of guest workers will spark political unrest and abandonment of traditional Islamic customs.

Migration by Asians nearly a century ago is producing contemporary problems in several countries.

(104)

The argument of anti-immigrant politicians is seductive to many voters in Western Europe, as well as the United States. Such arguments have little scientific basis, and in a culturally diverse world these arguments have racist overtones.

Key Issue 4. Why Do People Migrate within a Country?
- **Migration between regions of a country**
- **Migration within one region**

Internal migration for most people is less disruptive than international migration. Two main types of internal migration are interregional and intraregional. The principal type of interregional migration is between rural and urban areas, while the main type of intraregional migration is from older cities to suburbs.

Migration between Regions of a Country

In the United States, interregional migration was more prevalent in the past, when most people were farmers.

Migration between Regions Within the United States

The most famous example of large-scale internal migration is the opening of the American West.

Changing Center of Population. The population center is the average location of everyone in the country, the "center of population gravity."

The changing location of the population center graphically demonstrates the march of the American people across the North American continent over the past 200 years. In 1790, the population center was located in Chesapeake Bay, east of Baltimore, Maryland.

(105)

Early Settlement in the Interior. By 1830 the center of population moved west of Moorefield, West Virginia. After 1830 the U.S. population center moved west more rapidly, to just west of Cincinnati, Ohio, in 1880. The population center shifted west rapidly because most western pioneers during the mid-nineteenth century passed through the interior of the country on their way to California. For much of the nineteenth century the continuous westward advance of settlement stopped at the 98th meridian. The interior of the country confronted early settlers with a physical environment that was unsuited to familiar agricultural practices. Maps at the time labeled the region west of the 98th meridian as the Great American Desert.

Settlement of the Great Plains. The U.S. population center continued to migrate westward at a much slower pace after 1880, . . . in part because large-scale migration to the East Coast . . . offset some of the migration . . . to the U.S. West. The westward movement of the U.S. population center also slowed after 1880 because people began to fill in the area between the 98th meridian and California.

(106)
Between 1950 and 1980 the population center moved west faster. In 1980 (it) jumped west of the Mississippi River. By 2000 the center had migrated into south-central Missouri.

Recent Growth of the South. During the 1990s, for the first time more Americans migrated out of the West than into the West. The population center since 1980 has moved southward more sharply.

Americans are immigrating into the South primarily for job opportunities. People also migrate to the South for environmental reasons.

(107)
Interregional migration has slowed considerably in the United States into the twenty-first century. Net migration between each pair of regions is now close to zero.

Migration between Regions in Other Countries
As in the United States, long-distance interregional migration has been an important means of opening new regions for economic development in other large countries.

Russia. Soviet policy encouraged factory construction near raw materials rather than near existing population concentrations (see Chapter 11). The collapse of the Soviet Union ended policies that encouraged interregional migration. In the transition to a market-based economy, Russian government officials no longer dictate "optimal" locations for factories.

Brazil. Most Brazilians live in a string of large cities near the Atlantic Coast. To increase the attractiveness of the interior, the government moved its capital in 1960 from Rio to a newly built city called Brasília.

Indonesia. Since 1969 the Indonesian government has paid for the migration of more than 5 million people, primarily from the island of Java, where nearly two-thirds of its people live, to less populated islands.

(108)
The number of participants has declined in recent years, primarily because of environmental concerns.

Europe. Throughout Western Europe . . . the regions with net immigration are also the ones with the highest per capita incomes. Even countries that occupy relatively small land areas have important interregional migration trends. Regional differences in economic conditions within European countries may become greater with increased integration of the continent's economy.

India. Indians require a permit to migrate—or even to visit—the State of Assam. The restrictions, which date from the British colonial era, are designed to protect the ethnic identity of Assamese.

Migration within One Region

While interregional migration attracts considerable attention, far more people move within the same region, which is *intraregional* migration. Less than 5 percent of the world's people lived in urban areas in 1800, compared to nearly half today.

Migration from Rural to Urban Areas
Urbanization began in the 1800s in the countries of Europe and North America that were undergoing rapid industrial development.

Migration from rural to urban areas has skyrocketed in recent years in the less developed countries of Africa, Asia, and Latin America.

Migration from Urban to Suburban Areas
In more developed countries, most intraregional migration is from central cities out to the suburbs. As a result of suburbanization, the territory occupied by urban areas has rapidly expanded (see Chapter 13).

(109)
Migration from Metropolitan to Nonmetropolitan Areas
During the late twentieth century, the more developed countries of North America and Western Europe witnessed a new trend. More people in these regions immigrated into rural areas than emigrated out of them. Net migration from urban to rural areas is called **counterurbanization**.

Most counterurbanization represents genuine migration from cities and suburbs to small towns and rural communities. Like suburbanization, people move from urban to rural areas for lifestyle reasons. Many migrants from urban to rural areas are retired people.

Counterurbanization has stopped in the United States because of poor economic conditions in some rural areas. Future migration trends are unpredictable in more developed countries, because future economic conditions are difficult to forecast.

Key Terms
Brain drain (p.99)
Chain migration (p.98)
Circulation (p.85)
Counterurbanization (p.109)
Emigration (p.85)
Floodplain (p.87)
Forced migration (p.88)
Guest workers (p.99)
Immigration (p.85)
Internal migration (p.88)
International migration (p.88)
Interregional migration (p.88)

Intervening obstacle (p.88)
Intraregional migration (p.88)
Migration (p.85)
Migration transition (p.89)
Mobility (p.85)
Net migration (p.85)
Pull factor (p.85)
Push factor (p.85)
Quotas (p.99)
Refugees (p.86)
Undocumented immigrants (p.96)
Voluntary migration (p.88)

Chapter 4. Folk and Popular Culture

People living in other locations often have extremely different social customs. Geographers ask why such differences exist and how social customs are related to the cultural landscape.

Key Issues
1. Where do folk and popular cultures originate and diffuse?
2. Why is folk culture clustered?
3. Why is popular culture widely distributed?
4. Why does globalization of popular culture cause problems?

(117)

In Chapter 1, *culture* was shown to combine three things: values, material artifacts, and political institutions. This chapter deals with the material artifacts of culture, the visible objects that a group possesses and leaves behind for the future. This chapter examines two facets of material culture. First is . . . survival activities. Second is . . . leisure activities—the arts and recreation. Culture can be distinguished from habit and custom.

A **habit** is a repetitive act that a particular *individual* performs. A custom is a repetitive act of a group. A collection of social customs produces a group's material culture. Material culture falls into two basic categories that differ according to scale: folk and popular. **Folk culture** is traditionally practiced primarily by small, homogeneous groups living in isolated rural areas. **Popular culture** is found in large, heterogeneous societies. Landscapes dominated by a collection of folk customs change relatively little over time. In contrast, popular culture is based on rapid simultaneous global **connections**. Thus, folk culture is more likely to vary from place to place at a given time, whereas popular culture is more likely to vary from time to time at a given place.

In Earth's *globalization*, popular culture is becoming more dominant, threatening the survival of unique folk cultures. The disappearance of local folk customs reduces local diversity in the world and the intellectual stimulation that arises from differences in background. The dominance of popular culture can also threaten the quality of the environment.

Key Issue 1. Where Do Folk and Popular Cultures Originate and Diffuse?
- **Origin of folk and popular cultures**
- **Diffusion of folk and popular cultures**

Two basic factors help explain the spatial differences between popular and folk cultures: the process of origin and the pattern of diffusion.

(118)
Origin of Folk and Popular Cultures
A social custom originates at a hearth, a center of innovation. Folk customs often have anonymous hearths. They may also have multiple hearths. Popular culture is most often a product of the economically more developed countries. Industrial technology permits the uniform reproduction of objects in large quantities.

Origin of Folk Music
Music exemplifies the differences in the origins of folk and popular culture. Folk songs tell a story or convey information about daily activities such as farming, life-cycle events (birth, death, and marriage), or mysterious events such as storms and earthquakes. Folk customs may have multiple origins owing to non-communication among groups in different places. Within the Upper South, geographer George Carney identified four major hearths of country music during the late nineteenth and early twentieth centuries.

Origin of Popular Music

In contrast to folk music, popular music is written by specific individuals for the purpose of being sold to a large number of people. Popular music as we know it today originated around 1900. To provide songs for music halls and vaudeville, a music industry was developed in New York, along 28th Street between Fifth Avenue and Broadway, a district that became known as Tin Pan Alley.

(119)

The diffusion of American popular music worldwide began in earnest during World War II, when the Armed Forces Radio Network broadcast music to American soldiers. English became the international language for popular music.

Hip hop is a more recent form of popular music that also originated in New York. Whereas Tin Pan Alley originated in Manhattan office buildings, hip hop originated in the late 1970s in the South Bronx. Hip hop demonstrates well the interplay between globalization and local diversity that is a prominent theme of this book. Lyrics make local references and represent a distinctive hometown scene. At the same time, hip hop has diffused rapidly around the world through instruments of globalization.

(120)
Diffusion of Folk and Popular Cultures
The broadcasting of American popular music on Armed Forces radio illustrates the difference in diffusion of folk and popular cultures. The spread of popular culture typically follows the process of hierarchical diffusion from hearths or nodes of innovation. In contrast, folk culture is transmitted . . . primarily through migration, . . . relocation diffusion.

The Amish: Relocation Diffusion of Folk Culture
Amish customs illustrate how relocation diffusion distributes folk culture. Amish folk culture remains visible on the landscape in at least 17 states. In Europe the Amish did not develop distinctive language, clothing, or farming practices and gradually merged with various Mennonite church groups. Several hundred Amish families migrated to North America in two waves.

Living in rural and frontier settlements relatively isolated from other groups, Amish communities retained their traditional customs, even as other European immigrants to the United States adopted new ones.

(121)
Sports: Hierarchical Diffusion of Popular Culture
In contrast with the diffusion of folk customs, organized sports provide examples of how popular culture is diffused. Many sports originated as isolated folk customs and were diffused like other folk culture, through the migration of individuals. The contemporary diffusion of organized sports, however, displays the characteristics of popular culture.

Folk Culture Origin of Soccer. Soccer is the world's most popular sport (it is called football outside North America). Its origin is obscure.

Early football games resembled mob scenes. In the twelfth century the rules became standardized. Because football disrupted village life, King Henry II banned the game from England in the late twelfth century. It was not legalized again until 1603 by King James I. At this point, football was an English folk custom rather than a global popular custom.

Globalization of Soccer. The transformation of football from an English folk custom to global popular culture began in the 1800s. Sport became a subject that was taught in school. Increasing leisure time permitted people not only to view sporting events but to participate in them. With higher incomes, spectators paid to see first-class events.

(122)

35

Football was first played in continental Europe in the late 1870s by Dutch students who had been in Britain. British citizens further diffused the game throughout the worldwide British Empire. In the twentieth century, soccer, like other sports, was further diffused by new communication systems, especially radio and television. Although soccer was also exported to the United States, it never gained the popularity it won in Europe and Latin America.

Sports in Popular Culture. Each country has its own preferred sports. Cricket is popular primarily in Britain and former British colonies. Ice hockey prevails, logically, in colder climates. The most popular sports in China are martial arts, known as wushu, including archery, fencing, wrestling, and boxing. Baseball . . . became popular in Japan after it was introduced by American soldiers.

Lacrosse is a sport played primarily in Ontario, Canada, and a few eastern U.S. cities, especially Baltimore and New York. It has also fostered cultural identity among the Iroquois Confederation of Six Nations. In recent years, the International Lacrosse Federation has invited the Iroquois nation to participate in the Lacrosse World Championships.

Despite the diversity in distribution of sports across Earth's surface and the anonymous origin of some games, organized spectator sports today are part of popular culture.

Key Issue 2. Why Is Folk Culture Clustered?
- **Isolation promotes cultural diversity**
- **Influence of physical environment**

Folk culture typically has unknown or multiple origins among groups living in relative isolation. A combination of physical and cultural factors influences the distinctive distributions of folk culture.

Isolation Promotes Cultural Diversity
Folk customs observed at a point in time vary widely from one place to another, even among nearby places.

Himalayan Art
In a study of artistic customs in the Himalaya Mountains, geographers P. Karan and Cotton Mather demonstrate that distinctive views of the physical environment emerge among neighboring cultural groups that are isolated.

(123)
These groups display similar uniqueness in their dance, music, architecture, and crafts.

Influence of the Physical Environment
People respond to their environment, but the environment is only one of several controls over social customs. Folk societies are particularly responsive to the environment because of their low level of technology and the prevailing agricultural economy. Yet folk culture may ignore the environment. Broad differences in folk culture arise in part from physical conditions and . . . these conditions produce varied customs. Two necessities of daily life—food and shelter—demonstrate the influence of cultural values and the environment on development of unique folk culture.

Distinctive Food Preferences
Folk food habits derive from the environment.

(124)
For example, rice demands a milder, moist climate, while wheat thrives in colder, drier regions. People adapt their food preferences to conditions in the environment. A good example is soybeans. In the raw state they are toxic and indigestible. Lengthy cooking renders . . . (soybeans) edible, but fuel is scarce in Asia. Asians derive . . . foods from soybeans that do not require extensive cooking. In Europe, traditional preferences for quick-frying foods in Italy resulted in part from fuel shortages. In

Northern Europe, an abundant wood supply encouraged the slow stewing and roasting of foods over fires, which also provided home heat in the colder climate.

Food Diversity in Transylvania. Food customs are inevitably affected by the availability of products, but people do not simply eat what is available in their particular environment. In Transylvania, currently part of Romania, food preferences distinguish among groups who have long lived in close proximity. Soup, the food consumed by poorer people, shows the distinctive traditions of the neighboring cultural groups in Transylvania.

Long after dress, manners, and speech have become indistinguishable from those of the majority, old food habits often continue as the last vestige of traditional folk customs.

Food Attractions and Taboos. According to many folk customs, everything in nature carries a signature, or distinctive characteristic, based on its appearance and natural properties. Certain foods are eaten because their natural properties are perceived to enhance qualities considered desirable by the society, such as strength, fierceness, or lovemaking ability. People refuse to eat particular plants or animals that are thought to embody negative forces in the environment. Such a restriction on behavior imposed by social custom is a **taboo**. Other social customs, such as sexual practices, carry prohibitions, but taboos are especially strong in the area of food.

(125)
Hindu taboos against consuming cows can also be explained partly for environmental reasons. A large supply of oxen must be maintained in India, because every field has to be plowed at approximately the same time: when the monsoon rains arrive.

But the taboo against consumption of meat among many people, including Muslims, Hindus, and Jews, cannot be explained primarily by environment factors. Social values must influence the choice of diet, because people in similar climates and with similar levels of income consume different foods.

Folk Housing. The house . . . is a product of both cultural tradition and natural conditions.
(126)
Distinctive Building Materials. The two most common building materials in the world are wood and brick. The choice of building materials is influenced both by social factors and by what is available from the environment.

Distinctive House Form and Orientation. Social groups may share building materials, but the distinctive form of their houses may result from customary beliefs or environmental factors. The form of houses in some societies might reflect religious values.

Beliefs govern the arrangement of household activities in a variety of Southeast Asian societies.

(127)
Housing and Environment. The form of housing is related in part to environmental as well as social conditions. Even in areas that share similar climates and available building materials, folk housing can vary, owing to minor differences in environmental features.

U.S. Folk House Forms
Older houses in the United States display local folk-culture traditions. The style of pioneer homes reflected whatever upscale style was prevailing at the place on the East Coast from which they migrated. In contrast, houses built in the United States during the past half century display popular culture influences.

Fred Kniffen identified three major hearths or nodes of folk house forms in the United States: New England, Middle Atlantic, and Lower Chesapeake.

(128)

Today, such distinctions are relatively difficult to observe in the United States. Rapid communication and transportation systems provide people throughout the country with knowledge of alternative styles. Furthermore . . . houses are usually mass-produced by construction companies.

(129)

Key Issue 3. Why Is Popular Culture Widely Distributed?
- **Diffusion of popular housing, clothing, and food**
- **Role of television in diffusing popular culture**

Popular culture varies more in time than in place. It diffuses rapidly across Earth to locations with a variety of physical conditions.

Diffusion of Popular Housing, Clothing, and Food
Some regional differences in food, clothing, and shelter persist in more developed countries, but differences are much less than in the past.

Popular Housing Styles
Housing built in the United States since the 1940s demonstrates how popular customs vary more in time than in place.

In contrast with folk housing that is characteristic of the early 1800s, newer housing in the United States has been built to reflect rapidly changing fashion concerning the most suitable house form. In the years immediately after World War II . . . most U.S. houses were built in a *modern style*. Since the 1960s, styles that architects call *neo-eclectic* have predominated.

Modern House Styles (1945–1960). In the late 1940s and early 1950s, the dominant type was known as *minimal traditional*. The *ranch* house replaced minimal traditional as the dominant style of housing in the 1950s.

(130)

The *split-level* house was a popular variant of the ranch house between the 1950s and 1970s. The *contemporary* style was an especially popular choice between the 1950s and 1970s for architect-designed houses. The *shed* style, popular in the late 1960s, was characterized by high-pitched shed roofs.

Neo-eclectic House Styles (Since 1960). In the late 1960s, *neo-eclectic* styles became popular and by the 1970s had surpassed modern styles in vogue. The first popular neo-eclectic style was the *mansard* in the late 1960s and early 1970s. The *neo-Tudor* style, popular in the 1970s, was characterized by dominant, steep-pitched front-facing gables and half-timbered detailing. The *neo-French* style also appeared in the early 1970s.
The *neo-colonial* style, an adaptation of English colonial houses, has been continuously popular since the 1950s but never dominant.

Rapid Diffusion of Clothing Styles
Individual clothing habits reveal how popular culture can be distributed across the landscape with little regard for distinctive physical features. In the more developed countries . . . clothing habits generally reflect occupations rather than particular environments.

A second influence on clothing in MDCs is higher income. Improved communications have permitted the rapid diffusion of clothing styles from one region of Earth to another. Until recently, a year could elapse from the time an original dress was displayed to the time that inexpensive reproductions were available in the stores. Now the time lag is less than six weeks. The globalization of clothing styles has involved increasing awareness by North Americans and Europeans of the variety of folk costumes around the world. The continued use of folk costumes in some parts of the globe may persist not because of distinctive environmental conditions or traditional cultural values but to preserve past memories or to attract tourists.

Jeans. An important symbol of the diffusion of western popular culture is jeans, which became a prized possession for young people throughout the world. Locally made denim trousers are available throughout Europe and Asia for under $10, but "genuine" jeans made by Levi Strauss, priced at $50 to $100, are preferred as a status symbol.

(132)
Jeans became an obsession and a status symbol among youth in the former Soviet Union, when the Communist government prevented their import. The scarcity of high-quality jeans was just one of many consumer problems that were important motives in the dismantling of Communist governments in Eastern Europe around 1990.

Popular Food Customs
People in a country with a more developed economy are likely to have the income, time, and inclination to facilitate greater adoption of popular culture.
Alcohol and Fresh Produce. Consumption of large quantities of alcoholic beverages and snack foods are characteristic of the food customs of popular societies. Americans choose particular beverages or snacks in part on the basis of preference for what is produced, grown, or imported locally. However, cultural backgrounds also affect the amount and types of alcohol and snack foods consumed.

(133)
Geographers cannot explain all the regional variations in food preferences.

Wine Production. The spatial distribution of wine production demonstrates that the environment plays a role in the distribution of popular as well as folk food customs. Because of the unique product created by the distinctive soil and climate characteristics, the world's finest wines are most frequently identified by their place of origin. Although grapes can be grown in a wide variety of locations, wine distribution is based principally on cultural values, both historical and contemporary. Wine production is discouraged in regions of the world dominated by religions other than Christianity. The distribution of wine production shows that the diffusion of popular customs depends less on the distinctive environment of a location than on the presence of beliefs, institutions, and material traits conducive to accepting those customs.

Role of Television in Diffusing Popular Culture
Watching television is an especially significant popular custom for two reasons. First, it is the most popular leisure activity in more developed countries throughout the world. Second, television is the most important mechanism by which knowledge of popular culture, such as professional sports, is rapidly diffused across Earth.

Diffusion of Television
Inventors in a number of countries, including the United States, the United Kingdom, France, Germany, Japan, and the Soviet Union, simultaneously contributed to the development of television. The U.S. public first saw television in the 1930s. However, its diffusion was blocked for a number of years when broadcasting was curtailed or suspended entirely during World War II.

(134)
During the early 1950s, television sets were being sold in only 20 countries, and more than 85 percent of the world's 37 million sets were in the United States. By the early 1990s more than 180 countries had 900 million television sets, with less than one-fourth in the United States.

Currently, the level of television service falls into four categories. The first category consists of countries where nearly every household owns a TV set. A second category consists of countries in which ownership of a television is common but by no means universal, primarily Latin America . . . and the poorer European states. The third category consists of countries in which television exists but has not yet been widely diffused. Finally, about 30 countries, most of which are in Africa and Asia, have very few television sets.

Diffusion of the Internet
The diffusion of internet service is following the pattern established by television a generation earlier, and is likely to diffuse rapidly to other countries in the years ahead. Among less developed regions, Latin America and Asia are likely to expand internet hosts more rapidly than Africa.
(136)
Government Control of Television
In the United States most television stations are owned by private corporations. Some stations, however, are owned by local governments or other nonprofit organizations and are devoted to educational or noncommercial programs.

In most countries the government(s) . . . control TV stations to minimize the likelihood that programs hostile to current policies will be broadcast—in other words, they are censored. Operating costs are typically paid by the national government from tax revenues, although some government-controlled stations do sell air time to private advertisers. A number of Western European countries have transferred some government-controlled television stations to private companies.

Reduced Government Control. In the past, many governments viewed television as an important tool for fostering cultural integration. In recent years, changing technology—especially the diffusion of small satellite dishes—has made television a force for political change rather than stability.

(137)
Governments have had little success in shutting down satellite technology. The diffusion of small satellite dishes hastened the collapse of Communist governments in Eastern Europe during the late 1980s. Facsimile machines, portable video recorders, and cellular telephones have also put chinks in government censorship.

Key Issue 4. Why Does Globalization of Popular Culture Cause Problems?
- **Threat to folk culture**
- **Environmental impact of popular culture**

The international diffusion of popular culture has led to two problems. First, the diffusion of popular culture may threaten the survival of traditional folk culture in many countries. Second, popular culture may be less responsive to the diversity of local environments and consequently may generate adverse environmental impacts.

Threat to Folk Culture
When people turn from folk to popular culture, they may also turn away from the society's traditional values.

Loss of Traditional Values
One example of the symbolic importance of folk culture is clothing. In African and Asian countries today, there is a contrast between the clothes of rural farmworkers and of urban business and government leaders. The Western business suit has been accepted as the uniform for business executives and bureaucrats around the world. Wearing clothes typical of MDCs is controversial in some Middle Eastern countries.

Change in Traditional Role of Women. The global diffusion of popular culture threatens the subservience of women to men that is embedded in many folk customs. The concepts of legal equality and availability of economic and social opportunities outside the home have become widely accepted in more developed countries, even where women in reality continue to suffer from discriminatory practices.

(138)
However, contact with popular culture also has brought negative impacts for women in less developed countries, such as an increase in prostitution. International prostitution is encouraged in (some) countries as a major source of foreign currency.

Threat of Foreign Media Imperialism
Leaders of some LDCs consider the dominance of popular customs by MDCs as a threat to their independence.

(139)
Leaders of many LDCs view the spread of television as a new method of economic and cultural imperialism on the part of the more developed countries, especially the United States.

Western Control of News Media. Less developed countries fear the effects of the newsgathering capability of the media even more than their entertainment function. Many African and Asian government officials criticize the Western concept of freedom of the press. They argue that the American news organizations reflect American values and do not provide a balanced, accurate view of other countries. In many regions of the world the only reliable and unbiased news accounts come from the BBC World Service shortwave radio newscasts.

Environmental Impact of Popular Culture
Popular culture is less likely than folk culture to be distributed with consideration for physical features.

Modifying Nature
Popular culture can significantly modify or control the environment.

(140)
It may be imposed on the environment rather than springing forth from it, as with many folk customs.

Diffusion of Golf. Golf courses, because of their large size (80 hectares, or 200 acres), provide a prominent example of imposing popular culture on the environment. Golf courses are designed partially in response to local physical conditions. Yet, like other popular customs, golf courses remake the environment.

Uniform Landscapes
The distribution of popular culture around the world tends to produce more uniform landscapes. In fact, promoters of popular culture want a uniform appearance to generate "product recognition" and greater consumption.

Fast-food Restaurants. The diffusion of fast-food restaurants is a good example of such uniformity.

(141)
The success of fast-food restaurants depends on large-scale mobility. Uniformity in the appearance of the landscape is promoted by a wide variety of other popular structures in North America, such as gas stations, supermarkets, and motels. These structures are designed so that both local residents and visitors immediately recognize the purpose of the building, even if not the name of the company.

Global Diffusion of Uniform Landscapes. Diffusion of popular culture across Earth is not confined to products that originate in North America. Japanese automobiles and electronics, for example, have diffused in recent years to the rest of the world, including North America.

Negative Environmental Impact

The diffusion of some popular customs can adversely impact environmental quality in two ways: depletion of scarce natural resources and pollution of the landscape.

Increased Demand for Natural Resources. Diffusion of some popular customs increases demand for raw materials.

(142)

Increased demand for some products can strain the capacity of the environment. With a large percentage of the world's population undernourished, some question . . . inefficient use of grain to feed animals for eventual human consumption.

Pollution. Popular culture also can pollute the environment. Folk culture, like popular culture, can also cause environmental damage, especially when natural processes are ignored. A widespread belief exists that indigenous peoples of the Western Hemisphere practiced more "natural," ecologically sensitive agriculture before the arrival of Columbus and other Europeans. Geographers increasingly question this. Very high rates of soil erosion have been documented in Central America from the practice of folk culture.

The more developed societies that produce endless supplies for popular culture have created the technological capacity both to create large-scale environmental damage and to control it. However, a commitment of time and money must be made to control the damage. Adverse environmental impact of popular culture is further examined in Chapter 14.

Key Terms
Custom (p.117)
Folk culture (p.117)
Habit (p.117)
Popular culture (p.117)
Taboo (p.125)

Chapter 5. Language

Earth's heterogeneous collection of languages is one of its most obvious examples of cultural diversity. Estimates of distinct languages in the world range from 2,000 to 4,000. Including the 10 largest ones, altogether only about 100 languages are spoken by at least 5 million people, and another 70 by between 2 million and 5 million people.

Key Issues
1. Where are English-language speakers distributed?
2. Why is English related to other languages?
3. Where are other language families distributed?
4. Why do people preserve local languages?

(149)

Language is a system of communication through speech. Many languages also have a **literary tradition**, or a system of written communication. The lack of written record makes it difficult to document the distribution of many languages. Countries designate at least one language as their **official language**. A country with more than one official language may require all public documents to be in all languages.

We start our study of the geographic elements of cultural values with language in part because it is the means through which other cultural values, such as religion and ethnicity, are communicated. The study of language follows logically from migration, because the contemporary distribution of languages around the world results largely from past migrations of peoples. The final section of the chapter discusses contradictory trends of *scale* in language. On the one hand, English has achieved an unprecedented *globalization*. On the other hand, people are trying to preserve *local diversity* in language. The global distribution of languages results from a combination of two geographic processes—interaction and isolation. The Indo-European language family developed as a result of migration and subsequent isolation of people that can only be reconstructed through linguistic and archaeological theories.

Key Issue 1. Where Are English-Language Speakers Distributed?
- **Origin and diffusion of English**
- **Dialects of English**

A language originates at a particular place and diffuses to other locations through the migration of its speakers.

Origin and Diffusion of English
English is spoken fluently by one-half billion people, more than any language except for Mandarin. Whereas nearly all Mandarin speakers are clustered in one country—China—English speakers are distributed around the world.

English Colonies
The contemporary distribution of English speakers around the world exists because the people of England migrated with their language when they established colonies during the past four centuries.

(150)

English first diffused west from England to North America in the seventeenth century. Similarly, the British took control of Ireland in the seventeenth century, South Asia in the mid-eighteenth century, the South Pacific in the late eighteenth and early nineteenth centuries, and southern Africa in the late nineteenth century. More recently, the United States has been responsible for diffusing English to several places.

Origin of English in England

The British Isles had been inhabited for thousands of years, but we know nothing of their early languages, until tribes called the Celts arrived around 2000 B.C. Then, around A.D. 450, tribes from mainland Europe invaded, pushing the Celts into the remote northern and western parts.

German Invasion. The invading tribes were the Angles, Jutes, and Saxons. All three were Germanic tribes. England comes from Angles' land. In Old English, Angles was spelled Engles. The (Angles) came from a corner, or angle, of Germany known as Schleswig-Holstein.

(151)
Other peoples subsequently invaded England and added . . . to the basic English. Although the Vikings failed in their attempt to conquer the islands, many remained in the country to enrich the language with new words.

Norman Invasion. English is . . . different from German . . . because England was conquered by the Normans in 1066. The Normans, who came from present-day Normandy in France, spoke French, which they established as England's official language for the next 150 years. The majority of the people continued to speak English.

In 1204 . . . England lost control of Normandy and entered a long period of conflict with France. Parliament enacted the Statute of Pleading in 1362 to change the official language of court business from French to English.

(152)
During the 300-year period that French was the official language of England, the Germanic language used by the common people and the French used by the leaders mingled to form a new language.

Dialects of English
A **dialect** is a regional variation of a language distinguished by distinctive vocabulary, spelling, and pronunciation. English has an especially large number of dialects. One particular dialect of English, the one associated with upper-class Britons living in the London area, is recognized in much of the English-speaking world as the standard form of British speech, . . . known as **British Received Pronunciation (BRP).**

Dialects in England
English originated with three invading groups . . . who settled in different parts of Britain. The language each spoke was the basis of distinct regional dialects of Old English. Following the Norman invasion of 1066, . . . by the time English again became the country's dominant language, five major regional dialects had emerged.

From this large collection of local dialects, one eventually emerged as the standard language . . . the dialect used by upper-class residents in the capital city of London and the two important university cities of Cambridge and Oxford . . . first encouraged by the introduction of the printing press to England in 1476.

(153)
Grammar books and dictionaries printed in the eighteenth century established rules for spelling and grammar that were based on the London dialect. Strong regional differences persist . . . in the United Kingdom, especially in rural areas. Several dozen dialects . . . can be grouped into three main ones: Northern, Midland, and Southern. Further, distinctive southwestern and southeastern accents occur within the Southern dialect.

Differences between British and American English
The earliest colonists were most responsible for the dominant language patterns that exist today in the English-speaking part of the Western Hemisphere.

Differences in Vocabulary and Spelling. English in the United States and England evolved independently during the eighteenth and nineteenth centuries. U.S. English differs from that of England in three significant ways: vocabulary, spelling, and pronunciation. The vocabulary is different . . . because settlers in America encountered many new objects and experiences, . . . which were given names borrowed from Native Americans. As new inventions appeared, they acquired different names on either side of the Atlantic. Spelling diverged . . . because of a strong national feeling in the United States for an independent identity. Noah Webster, the creator of the first comprehensive American dictionary and grammar books, was not just a documenter of usage, he had an agenda. Webster argued that spelling and grammar reforms would help establish a national language, reduce cultural dependence on England, and inspire national pride.

(154)

Differences in Pronunciation. Differences in pronunciation between British and U.S. speakers are immediately recognizable. Interaction between the two groups was largely confined to exchange of letters and other printed matter rather than direct speech. Surprisingly, pronunciation has changed more in England than in the United States. People in the United States do not speak "proper" English because when the colonists left England, "proper" English was not what it is today.

Dialects in the United States
Major differences in U.S. dialects originated because of differences in dialects among the original settlers.

Settlement in the East. The original American settlements . . . can be grouped into three areas: New England, Middle Atlantic, and Southeastern. Two-thirds of the New England colonists were Puritans from East Anglia in southeastern England. About half of the southeastern settlers came from southeast England, although they represented a diversity of social-class backgrounds. The immigrants to the Middle Atlantic colonies were more diverse . . . because most of the settlers came from the north rather than the south of England or from other countries.

(155)

Current Dialect Differences in the East. Today, major dialect differences within the United States continue to exist, primarily on the East Coast. Every word that is not used nationally has some geographic extent . . . and therefore has boundaries . . . known as an isogloss. Two important isoglosses separate the eastern United States into three major dialect regions, known as Northern, Midland, and Southern. Some words are commonly used within one of the three major dialect areas but rarely in the other two. In most instances, these words relate to rural life, food, and objects from daily activities. Many words that were once regionally distinctive are now national in distribution. Mass media, especially television and radio, influence the adoption of the same words throughout the country.

Pronunciation Differences. Regional pronunciation differences are more familiar to us than word differences, although it is harder to draw precise isoglosses for them. The New England accent is well known for dropping the /r/ sound, . . . shared with speakers from the south of England. Residents of Boston . . . maintained especially close ties to the important ports of southern England. Compared to other colonists, New Englanders received more exposure to changes in pronunciation that occurred in Britain during the eighteenth century.

(156)

The mobility of Americans has been a major reason for the relatively uniform language that exists throughout much of the West.

Key Issue 2. Why Is English Related to Other Languages?
- **Indo-European branches**
- **Origin and diffusion of Indo-European**

45

English is part of the Indo-European language family. A language family is a collection of languages related through a common ancestor that existed long before recorded history.

Indo-European Branches
Within a language family, a **language branch** is a collection of languages related through a common ancestor that existed several thousand years ago.

(157)
Indo-European is divided into eight branches. Four of the branches—Indo-Iranian, Romance, Germanic, and Balto-Slavic—are spoken by large numbers of people. The four less extensively used Indo-European language branches are Albanian, Armenian, Greek, and Celtic.

Germanic Branch of Indo-European
A **language group** is a collection of languages within a branch that share a common origin in the relatively recent past. English and German are both languages in the West Germanic group. West Germanic is further divided into High Germanic and Low Germanic subgroups, so named because they are found in high and low elevations within present-day Germany. High German, spoken in the southern mountains of Germany, is the basis for the modern standard German language. English is classified in the Low Germanic subgroup. The Germanic language branch also includes North Germanic languages, spoken in Scandinavia. The four Scandinavian languages—Swedish, Danish, Norwegian, and Icelandic—all derive from Old Norse.

Indo-Iranian Branch of Indo-European
The branch of the Indo-European language family with the most speakers is Indo-Iranian, . . . more than 100 individual languages . . . divided into an eastern group (Indic) and a western group (Iranian).

Indic (Eastern) Group of Indo-Iranian Language Branch. The most widely used languages in India, as well as in the neighboring countries of Pakistan and Bangladesh, belong . . . to the Indic group of the Indo-Iranian branch of Indo-European. Approximately one-third of Indians, mostly in the north, use an Indic language called Hindi. Hindi is spoken many different ways—and therefore could be regarded as a collection of many individual languages—but there is only one official way to write the language, using a script called Devanagari.

(158)
Pakistan's principal language, Urdu, is spoken very much like Hindi but is written with the Arabic alphabet, a legacy of the fact that most Pakistanis are Muslims, and their holiest book (the Quran) is written in Arabic. Hindi, originally a variety of Hindustani spoken in the area of New Delhi, grew into a national language in the nineteenth century when the British encouraged its use in government.

India has four important language families: Indo-European (predominantly in the north), Dravidian (in the south), Sino-Tibetan (in the northeast), and Austro-Asiatic (in the central and eastern highlands). India's constitution as amended recognizes 18 official languages, including 13 Indo-European, 4 Dravidian, and 1 Sino-Tibetan language (Manipuri). As the language of India's former colonial ruler, English has an "associate" status, even though only 1 percent of the Indian population can speak it.

Iranian (Western) Group of Indo-Iranian Language Branch. Indo-Iranian languages . . . spoken in Iran and neighboring countries . . . form a separate group from Indic.

(159)
The major Iranian group languages include Persian (sometimes called Farsi) in Iran, Pathan in eastern Afghanistan and western Pakistan, and Kurdish, used by the Kurds of western Iran, northern Iraq, and eastern Turkey. These languages are written in the Arabic alphabet.

Balto-Slavic Branch of Indo-European

Slavic was once a single language, but differences developed in the seventh century A.D. when several groups of Slavs migrated from Asia to different areas of Eastern Europe.

East Slavic and Baltic Groups of Balto-Slavic Language Branch. The most widely used Slavic languages are the eastern ones, primarily Russian. With the demise of the Soviet Union, the newly independent republics adopted official languages other than Russian, although Russian remains the language for communications among officials in the countries that were formerly part of the Soviet Union. After Russian, Ukrainian and Belarusian (sometimes written Byelorussian) are the two most important East Slavic languages. The desire to use languages other than Russian was a major drive in the Soviet Union breakup a decade ago.

West and South Slavic Groups of Balto-Slavic Language Branch. The most spoken West Slavic language is Polish, followed by Czech and Slovak. The latter two are quite similar, and speakers of one can understand the other. The two most important South Slavic languages are Serbo-Croatian and Bulgarian. Although Serbs and Croats speak the same language, they use different alphabets. Slovene is the official language of Slovenia, while Macedonian is used in the former Yugoslav republic of Macedonia.

In general, differences among all Slavic languages are relatively small. However, because language is a major element in a people's cultural identity, relatively small differences among Slavic as well as other languages are being preserved and even accentuated in recent independence movements. Since Bosnia and Croatia broke away from Serb-dominated Yugoslavia in the early 1990s, regional differences within Serbo-Croatian have increased.

(160)
Romance Branch of Indo-European
The Romance language branch evolved from the Latin language spoken by the Romans 2,000 years ago. The four most widely used contemporary Romance languages are Spanish, Portuguese, French, and Italian. The physical boundaries such as mountains are strong intervening obstacles, creating barriers to communication between people living on opposite sides.

The fifth most important Romance language, Romanian, is the principal language of Romania and Moldova. Two other official Romance languages are Romansh and Catalán. Sardinian—a mixture of Italian, Spanish, and Arabic—once was the official language of the Mediterranean island of Sardinia. In addition to these official languages, several other Romance languages have individual literary traditions. In Italy, Ladin (not Latin) . . . and Friulian . . . (along with the official Romansh) are dialects of Rhaeto-Romanic. Ladino—a mixture of Spanish, Greek, Turkish, and Hebrew—is spoken by 140,000 Sephardic Jews, most of whom now live in Israel.

Origin and Diffusion of Romance Languages. As the conquering Roman armies occupied the provinces of this vast empire, they brought the Latin language with them.

(161)
The languages spoken by the natives of the provinces were either extinguished or suppressed. Latin used in each province was based on that spoken by the Roman army at the time of occupation. Each province also integrated words spoken in the area. The Latin that people in the provinces learned was not the standard literary form but a spoken form, known as **Vulgar Latin**, from the Latin word referring to "the masses" of the populace.

By the eighth century, regions of the former empire had been isolated from each other long enough for distinct languages to evolve. Latin persisted in parts of the former empire. People in some areas reverted to former languages, while others adopted the languages of conquering groups from the north and east, which spoke Germanic and Slavic.

Romance Language Dialects. Distinct Romance languages did not suddenly appear. They evolved over time. The creation of standard national languages, such as French and Spanish, was relatively recent. The dialect of the Île-de-France region, known as Francien, became the standard form of French because the region included Paris. The most important surviving dialect difference within France is between the north and the south. The northern dialect, **langue d'oïl** and the southern **langue d'òc** . . . provide insight into how languages evolve. These terms derive from different ways in which the word for "yes" was said. Spain, like France, contained many dialects during the Middle Ages. In the fifteenth century, when the Kingdom of Castile and Léon merged with the Kingdom of Aragón, . . .Castilian became the official language for the entire country. Spanish and Portuguese have achieved worldwide importance because of the colonial activities of their European speakers. Approximately 90 percent of the speakers of these two languages live outside Europe. Spanish is the official language of 18 Latin American states, while Portuguese is spoken in Brazil. The division of Central and South America into Portuguese- and Spanish-speaking regions is the result of a 1493 decision by Pope Alexander VI. The Portuguese and Spanish languages spoken in the Western Hemisphere differ somewhat from their European versions.

(162)
Difficulties arise in determining whether two languages are distinct or whether they are merely two dialects of the same language.

A creole or creolized language is defined as a language that results from the mixing of the colonizer's language with the indigenous language. A creolized language forms when the colonized group . . . makes some changes, such as simplifying the grammar. The word creole derives from a word in several Romance languages for a slave who is born in the master's house.

Origin and Diffusion of Indo-European
If Germanic, Romance, Balto-Slavic, and Indo-Iranian . . . are all part of the same . . . language family, then they must be descended from a single common ancestral language. The existence of a single ancestor . . . cannot be proved with certainty, because it would have existed thousands of years before the invention of writing or recorded history. The evidence that Proto-Indo-European once existed is "internal." Individual Indo-European languages share common root words for winter and snow but not for ocean. Therefore, linguists conclude that original Proto-Indo-European speakers probably lived in a cold climate, or one that had a winter season, but did not come in contact with oceans.

(163)
One theory argues that language diffused primarily through warfare and conquest, while the other theory argues that the diffusion resulted from peaceful sharing of food. One influential hypothesis, espoused by Marija Gimbutas, is that the first Proto-Indo-European speakers were the Kurgan people, whose homeland was in the steppes near the border between present-day Russia and Kazakhstan. Between 3500 and 2500 B.C., Kurgan warriors, using their domesticated horses as weapons, conquered much of Europe and South Asia.

Not surprisingly, scholars disagree on where and when the first speakers of Proto-Indo-European lived. Archaeologist Colin Renfrew argues that they lived 2,000 years before the Kurgans, in eastern Anatolia, part of present-day Turkey.

(164)
The Indo-Iranian branch . . . originated either directly through migration from Anatolia, or indirectly by way of Russia north of the Black and Caspian seas. Renfrew argues that Indo-European diffused . . . with agricultural practices rather than by military conquest. After many generations of complete isolation, individual groups evolved increasingly distinct languages.

(165)
Key Issue 3. Where Are Other Language Families Distributed?

- **Classification of languages**
- **Distribution of language families**

Although several thousand languages are spoken, they can be organized logically into a small number of language families, . . . further divided into language branches and language groups.

Classification of Languages
About 50 percent of all people speak a language in the Indo-European family. About 20 percent speak a language in the Sino-Tibetan family.

(166)
About 5 percent each speak a language in one of these four families: Afro-Asiatic (in the Middle East); Austronesian (in Southeast Asia); Niger-Congo (in Africa); (and) Dravidian (in India).

Distribution of Language Families
Half the people in the world speak an Indo-European language. The second-largest family is Sino-Tibetan, spoken by nearly one-fourth of the world. Other major language families include Afro-Asiatic, Altaic, Austronesian, Japanese, and Niger-Congo.

Sino-Tibetan Family
The Sino-Tibetan family encompasses languages spoken in the People's Republic of China   . . . as well as several smaller countries in Southeast Asia.

Sinitic Branch. There is no single Chinese language. Spoken by approximately three-fourths of the Chinese people, Mandarin is by a wide margin the most used language in the world.

(167)
Other Sinitic branch languages are spoken by tens of millions of people in China. The Chinese government is imposing Mandarin countrywide. Unity is also fostered by a consistent written form for all Chinese languages. Although the words are pronounced differently in each language, they are written the same way. The structure of Chinese languages is quite different (from Indo-European). They are based on 420 one-syllable words.

(168)
This number far exceeds the possible one-syllable sounds that humans can make, so Chinese languages use each sound to denote more than one thing. The listener must infer the meaning from the context in the sentence and the tone of voice the speaker uses.

In addition, two one-syllable words can be combined. The other distinctive characteristic of the Chinese languages is the method of writing . . . with a collection of thousands of characters. Some . . . represent sounds. Most are ideograms, which represent ideas or concepts, not specific pronunciations.

Austro-Thai and Tibeto-Burman branches of Sino-Tibetan family. In addition to the Chinese languages included in the Sinitic branch, the Sino-Tibetan family includes two smaller branches, Austro-Thai and Tibeto-Burman.

Other East and Southeast Asian Language Families
Japanese and Korean both form distinctive language families. Chinese cultural traits have diffused into Japanese society, including the original form of writing the Japanese language. Japanese is written in part with Chinese ideograms, but it also uses two systems of phonetic symbols. Korean is usually classified as a separate language family. Korean is written not with ideograms but in a system known as hankul. In this system, each letter represents a sound. Austro-Asiatic, spoken by about 1 percent of the world's population, is based in Southeast Asia. Vietnamese (is) the most spoken tongue

of the . . . language family. The Vietnamese alphabet was devised in the seventh century by Roman Catholic missionaries.

Afro-Asiatic Language Family
The Afro-Asiatic—once referred to as the Semito-Hamitic—language family includes Arabic and Hebrew, as well as a number of languages spoken primarily in northern Africa and southwestern Asia.

(169)
Arabic is the major Afro-Asiatic language, an official language in two dozen countries of North Africa and southwestern Asia, from Morocco to the Arabian Peninsula.

Altaic and Uralic Language Families
The Altaic and Uralic language families were once thought to be linked as one family because the two display similar word formation, grammatical endings, and other structural elements. Recent studies, however, point to geographically distinct origins.

Altaic Languages. The Altaic languages are spoken across an 8,000-kilometer (5,000-mile) band of Asia between Turkey on the west and Mongolia and China on the east. Turkish (is) by far the most widely used . . . . Other Altaic languages with at least 1 million speakers include Azerbaijani, Bashkir, Chuvash, Kazakh, Kyrgyz, Mongolian, Tatar, Turkmen, Uighur, and Uzbek. With the dissolution of the Soviet Union in the early 1990s, Altaic languages became official in several newly independent countries. Enthusiasm for restoring languages long discouraged by the Soviet Union threatens the rights of minorities in these countries to speak other languages that are not officially recognized. Problems also persist because the boundaries of the countries do not coincide with the regions in which the speakers of the various languages are clustered.

Uralic Languages. Every European country is dominated by Indo-European speakers, except for three: Estonia, Finland, and Hungary. The Estonians, Finns, and Hungarians speak languages that belong to the Uralic family, . . . first used 7,000 years ago by people living in the Ural Mountains . . . . north of the Kurgan homeland.

African Language Families
No one knows the precise number of languages spoken in Africa, and scholars disagree on classifying the known ones into families. Nearly 1,000 distinct languages and several thousand named dialects have been documented. In northern Africa . . . an Arabic . . . dominates, although in a variety of dialects. Other Afro-Asiatic languages spoken by more than 5 million Africans include Amharic, Oromo, and Somali in the Horn of Africa, and Hausa in northern Nigeria.

(170)
In sub-Saharan Africa . . . languages grow far more complex.

Niger-Congo Language Family. More than 95 percent of the people in sub-Saharan Africa speak languages of the Niger-Congo family, which includes six branches with many hard to classify languages. The remaining 5 percent speak languages of the Khoisan or Nilo-Saharan families. Several million South Africans speak Indo-European languages. The largest branch of the Niger-Congo family is the Benue-Congo branch, and its most important language is Swahili. Its vocabulary has strong Arabic influences. Swahili is one of the few African languages with an extensive literature. **Nilo-Saharan Language Family.** Nilo-Saharan languages are spoken by a few million people in north-central Africa, immediately north of the Niger-Congo language region. Despite fewer speakers, the Nilo-Saharan family is divided into six branches.

Khoisan Language Family. The third important language family of sub-Saharan Africa—Khoisan—is concentrated in the southwest. Khoisan languages . . . use clicking sounds.

Austronesian Language Family. About 6 percent of the world's people speak an Austronesian language, once known as the Malay-Polynesian family. The most frequently used Austronesian language is Malay-Indonesian. The people of Madagascar speak Malagasy, which belongs to the Austronesian family, even though the island is separated by 3,000 kilometers (1,900 miles) from any other Austronesian-speaking country.

(171)

Nigeria: Conflict among Speakers of Different Languages. Africa's most populous country, Nigeria, displays problems that can arise from the presence of many speakers of many languages. Groups living in different regions of Nigeria have often battled. Nigeria reflects the problems that can arise when great cultural diversity—and therefore language diversity—is packed into a relatively small region.

Key Issue 4. Why Do People Preserve Local Languages?
- **Preserving language diversity**
- **Global dominance of English**

The distribution of a language is a measure of the fate of an ethnic group. As in other cultural traits, language displays the two competing geographic trends of globalization and local diversity.

Preserving Language Diversity
Thousands of languages are **extinct languages**, once in use—even in the recent past—but no longer spoken or read in daily activities by anyone in the world. The eastern Amazon region of Peru in the sixteenth century . . . (had) more than 500 languages. Only 57 survive today, half of which face extinction. Gothic was widely spoken . . . in Eastern and Northern Europe in the third century A.D.

(172)

The last speakers of Gothic lived in the Crimea in Russia in the sixteenth century. Many Gothic people switched to speaking the Latin language after their conversion to Christianity. Some endangered languages are being preserved. Nonetheless, linguists expect . . . that only about 300 languages are clearly safe from extinction.

Hebrew: Reviving Extinct Languages
Hebrew is a rare case of an extinct language that has been revived. Hebrew diminished in use in the fourth century B.C. and was thereafter retained only for Jewish religious services. When Israel was established . . . in 1948, Hebrew became one of the new country's two official languages, along with Arabic. The effort was initiated by Eliezer Ben-Yehuda, . . . credited with the invention of 4,000 new Hebrew words—related when possible to ancient ones—and the creation of the first modern Hebrew dictionary.

Celtic: Preserving Endangered Languages
Two thousand years ago Celtic languages were spoken in much of present-day Germany, France, and northern Italy, as well as in the British Isles. Today Celtic languages survive only in remoter parts of Scotland, Wales, and Ireland, and on the Brittany peninsula of France.

Celtic Groups. The Celtic language branch is divided into Goidelic (Gaelic) and Brythonic groups. Two Goidelic languages survive: Irish Gaelic and Scottish Gaelic. Only 75,000 people speak Irish Gaelic exclusively. In Scotland fewer than 80,000 of the people (2 percent) speak it. Over time, speakers of Brythonic (also called Cymric or Britannic) fled westward to Wales, southwestward to Cornwall, or southward across the English Channel to the Brittany peninsula of France. An estimated one-fourth of the people in Wales still use Welsh as their primary language, although all but a handful know English as well.

Cornish became extinct in 1777, with the death of the language's last known native speaker, Dolly Pentreath. An English historian recorded as much of her speech as possible so that future generations could study the Cornish language. In Brittany—like Cornwall, an isolated peninsula that juts out into the Atlantic Ocean—300,000 people still speak Breton. Only about 10,000 actually use Breton more than French. The Celtic languages declined because the Celts lost most of the territory they once controlled to speakers of other languages. In the 1300s the Irish were forbidden to speak their own language in the presence of their English masters.

(173)
Revival of Celtic Languages. Recent efforts have prevented the disappearance of Celtic languages. Britain's 1988 Education Act made Welsh language training a compulsory subject in all schools in Wales, and Welsh history and music have been added to the curriculum. The number of people fluent in Irish Gaelic has grown in recent years as well, especially among younger people. An Irish-language TV station began broadcasting in 1996. A couple of hundred people have now become fluent in the formerly extinct Cornish language, which was revived in the 1920s. Cornish is taught in grade schools and adult evening courses and is used in some church services. . . . However, a dispute has erupted over the proper way to spell Cornish words.

Multilingual States
Difficulties can arise at the boundary between two languages. The boundary between the Romance and Germanic branches runs through the middle of . . . Belgium and Switzerland. Belgium has had more difficulty than Switzerland in reconciling the interests of the different language speakers.

Belgium. Southern Belgians (known as Walloons) speak French, whereas northern Belgians (known as Flemings) speak a dialect of the Germanic language of Dutch, called Flemish.

(174)
Historically, the Walloons dominated Belgium's economy and politics. In response to pressure from Flemish speakers, Belgium was divided into two independent regions, Flanders and Wallonia. Belgium had difficulty fixing a precise boundary between Flemish and French speakers. During the late 1980s, this problem jailed one town's mayor and collapsed the national government.

Switzerland. In contrast, Switzerland peacefully exists with . . . four official languages: German, French, Italian, and Romansh. Swiss voters made Romansh an official language in a 1938 referendum, despite the small percentage who use the language. Other languages are used by nearly 10 percent of the Swiss population, mostly guest-worker immigrants.

(175)
Isolated Languages
An **isolated language** is a language unrelated to any other and therefore not attached to any language family. Isolated languages arise through lack of interaction with speakers of other languages.

A Pre-Indo-European Survivor: Basque. The best example of an isolated language in Europe is Basque. Basque is spoken by 1 million people in the Pyrenees Mountains.

An Unchanging Language: Icelandic. Unlike Basque, Icelandic is related to other languages. Icelandic's significance is that over the past thousand years it has changed less than any other in the Germanic branch.

(176)
Global Dominance of English
One of the most fundamental needs in a global society is a common language for communication. Increasingly in the modern world, the language of international communication is English.

When well-educated speakers of two different languages wish to communicate with each other in countries such as India or Nigeria, they frequently use English.

English: An Example of a Lingua Franca

A language of international communication . . . is known as a **lingua franca**. The term, which means *language of the Franks*, was originally applied by Arab traders during the Middle Ages to describe the language they used to communicate with Europeans, whom they called *Franks*. A group that learns English or another lingua franca may learn a simplified form, called a **pidgin language**. Two groups construct a pidgin language by learning a few of the grammar rules and words of a lingua franca, while mixing in some elements of their own languages. Other than English, modern lingua franca languages include Swahili in East Africa, Hindustani in South Asia, and Russian in the former Soviet Union.

Expansion Diffusion of English

In the past, a lingua franca achieved widespread distribution through migration and conquest. In recent centuries use of English spread around the world primarily through the British Empire. In contrast, the current growth in use of English around the world is a result not of military conquest, nor of migration by English-speaking people. Rather, the current growth in the use of English is an example of expansion diffusion, the spread of a trait through the snowballing effect of an idea rather than through the relocation of people. Unlike most examples of expansion diffusion . . . recent changes in English have percolated up from common usage and ethnic dialects rather than directed down to the masses by elite people. Examples include dialects spoken by African-Americans and residents of Appalachia. African-American slaves preserved a distinctive dialect in part to communicate in a code not understood by their white masters.

In the twentieth century . . . living in racially segregated neighborhoods within northern cities and attending segregated schools, many blacks preserved their distinctive dialect. That dialect has been termed Ebonics, a combination of ebony and phonics. The American Speech, Language and Hearing Association has classified Ebonics as a distinct dialect, with a recognized vocabulary, grammar, and word meaning. Natives of Appalachian communities, such as in rural West Virginia, also have a distinctive dialect.

(177)

Use of Ebonics is controversial within the African-American community. Similarly, speaking an Appalachian dialect produces both pride and problems.

Diffusion to Other Languages

English words have become increasingly integrated into other languages.

Franglais. The French are particularly upset with the increasing worldwide domination of English. French is an official language in 26 countries and for hundreds of years served as the lingua franca for international diplomats. The widespread use of English in the French language is called **franglais**, a combination of *français* and **anglais**, the French words for French and English.

(179)

Spanglish. English is diffusing into the Spanish language spoken by 17 million Hispanics in the United States, a process called **Spanglish**.

(180)

For example, shorts (pants) becomes chores, and vacuum cleaner becomes bacuncliner. Spanglish is a richer integration of English with Spanish than the mere borrowing of English words. New words have been invented in Spanglish that do not exist in English but would be useful if they did. Spanglish has become especially widespread in popular culture, such as song lyrics, television, and magazines aimed at young Hispanic women, but it has also been adopted by writers of serious literature.

Key Terms

British Received Pronunciation (BRP) (p.152)
Creole or creolized language (p.162)

Chapter 6. Religion

Most religious people pray for peace, but religious groups may not share the same vision of how peace will be achieved. Geographers see that the process by which one religion diffuses across the landscape may conflict with the distribution of others. Geographers also observe that religions are derived in part from elements of the physical environment, and that religions, in turn, modify the landscape.

(185)

Key Issues
1. Where are religions distributed?
2. Why do religions have different distributions?
3. Why do religions organize space in distinctive patterns?
4. Why do territorial conflicts arise among religious groups?

(187)

Religion interests geographers because it is essential for understanding how humans occupy Earth. Geographers, though, are not theologians, so they stay focused on those elements of religions that are geographically significant. Geographers study spatial *connections* in religion: the distinctive place of origin, . . . the extent of diffusion, . . . the processes by which religions diffused, . . . and practices and beliefs that lead some to have more widespread distributions.

Geographers find the tension in *scale* between *globalization* and *local diversity* especially acute in religion for a number of reasons. People care deeply about their religion; . . . some religions are . . . *designed* to appeal to people throughout the world, whereas other religions . . . appeal primarily . . . in geographically limited areas; religious values are important in . . . how people identify themselves, . . . (and) the . . . ways they organize the landscape; adopting a global religion usually requires turning away from a traditional local religion; and while migrants typically learn the language of the new location, they retain their religion.

This chapter starts by describing the distribution of major religions, then . . . explains why some religions have diffused widely, whereas others have not. The third section of the chapter discusses religion's strong imprint on the physical environment. Unfortunately, intense identification with one religion can lead adherents into conflicts discussed in the fourth key issue of the chapter.

Key Issue 1. Where Are Religions Distributed?
- **Universalizing religions**
- **Ethnic religions**

Universalizing religions attempt to be global, to appeal to all people. An **ethnic religion** appeals primarily to one group of people living in one place. About 60 percent of the world's population adheres to a universalizing religion, 25 percent to an ethnic religion, and 15 percent to no religion.

Universalizing Religions
The three main universalizing religions are Christianity, Islam, and Buddhism. Each is . . . divided into branches, denominations, and sects. A **branch** is a large and fundamental division within a religion. A **denomination** is a division of a branch that unites a number of local congregations. A **sect** is a relatively small group that has broken away from an established denomination.

Christianity. Christianity has about 2 billion adherents, far more than any other world religion, and has the most widespread distribution.

Branches of Christianity. Christianity has three major branches: Roman Catholic, Protestant, and Eastern Orthodox. Within Europe, Roman Catholicism is the dominant Christian branch in the southwest and east, Protestantism in the northwest, and Eastern Orthodoxy in the east and southeast.

The regions of Roman Catholic and Protestant majorities frequently have sharp boundaries, even when they run through the middle of countries.

(188)

The Eastern Orthodox branch of Christianity is a collection of 14 self-governing churches in Eastern Europe and the Middle East. More than 40 percent of all Eastern Orthodox Christians belong to . . . the Russian Orthodox Church, . . . established in the sixteenth century. Nine of the other 13 self-governing churches were established in the nineteenth or twentieth century. The largest of these 9, the Romanian church, includes 20 percent of all Eastern Orthodox Christians.

The remaining 4 of the 14 Eastern Orthodox churches— Constantinople, Alexandria, Antioch, and Jerusalem—trace their origins to the earliest days of Christianity. They have a combined membership of about 3 percent of all Eastern Orthodox Christians.

(189)

Christianity in the Western Hemisphere. The overwhelming percentage of people living in the Western Hemisphere—about 90 percent—are Christian. About 5 percent belong to other religions. Roman Catholics comprise 95 percent of Christians in Latin America, compared with 25 percent in North America. Within North America, Roman Catholics are clustered in the southwestern and northeastern United States and the Canadian province of Québec. Protestants comprise 40 percent of Christians in North America. The three largest Protestant denominations in the United States are Baptist, Methodist, and Pentecostal, followed by Lutheran, Latter-Day Saints, and Churches of Christ.

Membership in some Protestant churches varies by region of the United States. Baptists, for example, are highly clustered in the southeast, Lutherans in the upper Midwest, and Latter-Day Saints in Utah. Other Christian denominations are more evenly distributed around the country.

(190)

Smaller Branches of Christianity. Several other Christian churches developed independent of the three main branches. Many . . . were isolated . . . at an early point in the development of Christianity, partly because of differences in doctrine and partly as a result of Islamic control of intervening territory in Southwest Asia and North Africa. Two small Christian churches survive in northeast Africa: the Coptic Church of Egypt and the Ethiopian Church. The Armenian Church originated in Antioch, Syria, and was important in diffusing Christianity to South and East Asia between the seventh and thirteenth centuries. The Armenian Church, like other small sects, plays a significant role in regional conflicts. The Maronites, (clustered in Lebanon) are another example of a small Christian sect that plays a disproportionately prominent role in political unrest.
Islam. Islam, the religion of 1.2 billion people, is the predominant religion of the Middle East from North Africa to Central Asia. However, half of the world's Muslims live in four countries outside the Middle East: Indonesia, Pakistan, Bangladesh, and India.

(191)

Branches of Islam. Islam is divided into two important branches: Sunni (from the Arabic word for orthodox) and Shiite (from the Arabic word for sectarian, sometimes written Shia in English). Sunnis comprise 83 percent of Muslims and are the largest branch in most Muslim countries. Sixteen percent of Muslims are Shiites, clustered in a handful of countries.

Islam in North America and Europe. The Muslim population of North America and Europe has increased rapidly in recent years, mostly through immigration. In Europe, France has the largest Muslim population, a legacy of immigration from former colonies in North Africa. Islam also has a presence in the United States through the Nation of Islam, also known as Black Muslims, founded in Detroit in 1930 and led for more than 40 years by Elijah Muhammad, who called himself "the messenger of Allah." Since Muhammad's death, in 1975, his son Wallace D. Muhammad led the Black Muslims closer to the principles of orthodox Islam, and the organizations name was changed to the American Muslim Mission.

(192)
Buddhism. Buddhism, the third of the world's major universalizing religions, has 350 million adherents, especially in China and Southeast Asia.

Like the other two universalizing religions, Buddhism split into more than one branch. The three main branches are Mahayana, Theravada, and Tantrayana. An accurate count of Buddhists is especially difficult, because only a few people participate in Buddhist institutions. Buddhism . . . differs in significant respects from the Western concept of a formal religious system. Christianity and Islam both require exclusive adherence. Most Buddhists in China and Japan, in particular, believe at the same time in an ethnic religion.

Other Universalizing Religions. Sikhism and Bahá'í are the two universalizing religions other than Christianity, Islam, and Buddhism with the largest numbers of adherents. Sikhism's first guru (religious teacher or enlightener) was Nanak (A.D. 1469–1538), who lived in a village near the city of Lahore, in present-day Pakistan. The Bahá'í religion is even more recent than Sikhism. It grew out of the Bábi faith, which was founded in Shíráz, Iran, in 1844 by Siyyid 'Ali Muhammad, known as the Báb (Persian for gateway).

(193)
Ethnic Religions
The ethnic religion with by far the largest number of followers is Hinduism. With 900 million adherents, Hinduism is the world's third-largest religion, behind Christianity and Islam. Ethnic religions in Asia and Africa comprise most of the remainder.

Hinduism. Ethnic religions typically have much more clustered distributions than do universalizing religions. Ninety-seven percent of Hindus are concentrated in one country, India. Two percent are in the neighboring country of Nepal, and the remaining one percent are dispersed around the world.

The appropriate form of worship for any two individuals may not be the same. Hinduism does not have a central authority or a single holy book. The largest number of adherents—an estimated 70 percent— worships the god Vishnu, a loving god incarnated as Krishna. An estimated 25 percent adhere to . . . Siva, a protective and destructive god. Shaktism is a form of worship dedicated to the female consorts of Vishnu and Siva.

Other Ethnic Religions
Several hundred million people practice ethnic religions in East Asia, especially in China and Japan. Buddhism does not compete for adherents with Confucianism, Daoism, and other ethnic religions in China, because many Chinese accept the teachings of both universalizing and ethnic religions.

Confucianism. Confucius (551–479 B.C.) was a philosopher and teacher in the Chinese province of Lu. Confucianism prescribed a series of ethical principles for the orderly conduct of daily life in China.

Daoism (Taoism). Lao-Zi (604–531? B.C., also spelled Lao Tse), a contemporary of Confucius, organized Daoism. Daoists seek dao (or tao), which means the way or path. Dao cannot be comprehended by reason and knowledge, because not everything is knowable. Daoism split into many sects, some acting like secret societies, and followers embraced elements of magic.

Shintoism. Since ancient times, Shintoism has been the distinctive ethnic religion of Japan. Ancient Shintoists considered forces of nature to be divine, especially the Sun and Moon, as well as rivers, trees, rocks, mountains, and certain animals. Gradually, deceased emperors and other ancestors became more important deities for Shintoists than natural features. Shintoism still thrives in Japan, although no longer as the official state religion.

(194)
Judaism. About 6 million Jews live in the United States, 4 million in Israel, 2 million in former Soviet Union republics, . . . and 2 million elsewhere. The number of Jews living in the former Soviet Union has declined rapidly since the late 1980s, when emigration laws were liberalized.

Judaism plays a more substantial role in Western civilization than its number of adherents would suggest, because two of the three main universalizing religions—Christianity and Islam—find some of their roots in Judaism. The name Judaism derives from Judah, one of the patriarch Jacob's 12 sons; Israel is another biblical name for Jacob.

Ethnic African Religions. About 10 percent of Africans follow traditional ethnic religions, sometimes called animism. African animist religions are apparently based on monotheistic concepts, although below the supreme god there is a hierarchy of divinities, . . . assistants to god or personifications of natural phenomena, such as trees or rivers. Some atlases and textbooks persist in classifying Africa as predominantly animist, even though the actual percentage is small and declining. Africa is now nearly 50 percent Christian, and another 40 percent are Muslims. The growth in the two universalizing religions at the expense of ethnic religions reflects fundamental geographical differences between the two types of religions.

Key Issue 2. Why Do Religions Have Different Distributions?
- **Origin of religions**
- **Diffusion of religions**
- **Holy places**
- **The calendar**

We can identify several major geographical differences between universalizing and ethnic religions: locations where the religions originated, processes by which they diffused . . . to other regions, types of places . . . considered holy, calendar dates identified as important holidays, and attitudes toward modifying the physical environment.

(195)
Origin of Religions
Universalizing religions have precise places of origin, based on events in the life of a man. Ethnic religions have unknown or unclear origins, not tied to single historical individuals.

Origin of Universalizing Religions. Each of the three universalizing religions can be traced to the actions and teachings of a man who lived since the start of recorded history. Specific events also led to the division of the universalizing religions into branches.

Origin of Christianity. Christianity was founded upon the teachings of Jesus, who was born in Bethlehem between 8 and 4 B.C. and died on a cross in Jerusalem about A.D. 30.

Christians believe that Jesus died to atone for human sins, that he was raised from the dead by God, and that his Resurrection from the dead provides people with hope for salvation. Roman Catholics accept the teachings of the Bible, as well as the interpretation of those teachings by the Church hierarchy, headed by the Pope. Eastern Orthodoxy comprises the faith and practices of a collection of churches that arose in the Eastern part of the Roman Empire. The split between the Roman and Eastern churches dates to the fifth century, as a result of rivalry between the Pope of Rome and the Patriarchy of Constantinople. Protestantism originated with the principles of the Reformation in the sixteenth century.

Origin of Islam. Islam traces its origin to the same narrative as Judaism and Christianity. All three religions consider Adam to have been the first man and Abraham to have been one of his descendants. Jews and Christians trace their story through Abraham's original wife and son, Sarah and Isaac. Muslims trace their story through his second wife and son, Hagar and Ishmael. One of Ishmael's descendants, Muhammad, became the Prophet of Islam. Muhammad was born in Makkah about A.D. 570.

Differences between the two main branches—Shiites and Sunnis—go back to the earliest days of Islam and basically reflect disagreement over the line of succession in Islamic leadership.

(196)
Origin of Buddhism. The founder of Buddhism, Siddhartha Gautama, was born about 563 B.C. . . . in present-day Nepal, near the border with India. The son of a lord, he led a privileged existence sheltered from life's hardships. At age 29 Gautama left his palace . . . and lived in a forest for the next six years, thinking and experimenting with forms of meditation. Gautama emerged as the Buddha, the "awakened or enlightened one," and spent 45 years preaching his views across India. While the Theravadists emphasize Buddha's life of self-help and years of solitary introspection, Mahayanists emphasize Buddha's later years of teaching and helping others.

Origin of Other Universalizing Religions. Sikhism and Bahá'í were founded more recently than the three large universalizing religions. The founder of Sikhism, Guru Nanak, traveled widely through South Asia around 500 years ago preaching his new faith, and many people became his Sikhs, which is the Hindi word for disciples. When it was established in Iran during the nineteenth century, Bahá'í provoked strong opposition from Shiite Muslims. The Báb was executed in 1850, as were 20,000 of his followers.

Origin of Hinduism, an Ethnic Religion. Unlike the . . . universalizing religions, Hinduism did not originate with a specific founder. Hinduism existed prior to recorded history. Aryan tribes from Central Asia invaded India about 1400 B.C. and brought . . . their religion. Centuries of intermingling with the Dravidians already living in the area modified their religious beliefs.

Diffusion of Religions
The three universalizing religions diffused from specific hearths, or places of origin, to other regions of the world. In contrast, ethnic religions typically remain clustered in one location.

Diffusion of Universalizing Religions. The hearths . . . of the three largest universalizing religions . . . . are in Asia (Christianity and Islam in Southwest Asia, Buddhism in South Asia). Today these three . . . together have several billion adherents distributed across wide areas of the world.

(197)
Diffusion of Christianity. Christianity's diffusion has been rather clearly recorded . . . . Consequently, geographers can examine its diffusion by reconstructing patterns of communications, interaction, and migration. Chapter 1 identified two processes of diffusion—relocation (diffusion through migration) and expansion (diffusion through a snowballing effect)—and within expansion diffusion we distinguished between hierarchical (diffusion through key leaders) and contagious (widespread diffusion). Christianity diffused through a combination of all of these forms of diffusion. Christianity first diffused from its hearth in Palestine through relocation diffusion. **Missionaries** . . . carried the teachings of Jesus along the Roman Empire's protected sea routes and excellent road network . . . . People in commercial towns and military settlements that were directly linked by the communications network received the message first. Christianity (also) spread widely . . . through contagious diffusion—daily contact between believers in the towns and nonbelievers in the surrounding countryside. **Pagan**, the word for a follower of a polytheistic religion in ancient times, derives from the Latin word for *countryside*. The dominance of Christianity . . . was assured during the fourth century through hierarchical diffusion. Emperor Constantine . . . embraced it in A.D. 313, and Emperor Theodosius proclaimed it the empire's official religion in 380. In subsequent centuries Christianity further diffused into Eastern Europe through conversion of kings or other elite figures.

Migration and missionary activity . . . since . . . 1500 has extended Christianity to other regions, through permanent resettlement of Europeans, . . . by conversion of indigenous populations, and by intermarriage. In recent decades Christianity has further diffused to Africa, where it is now the most widely practiced religion. Latin Americans are predominantly Roman Catholic, colonized by the Spanish and Portuguese. Canada (except Québec) and the United States have Protestant majorities because colonists came primarily from Protestant England. Mormons, originating at Fayette, New York, then eventually migrated to the sparsely inhabited Salt Lake Valley in the present-day state of Utah.

(198)
Diffusion of Islam. Muhammad's successors organized followers into armies that extended the region of Muslim control over an extensive area of Africa, Asia, and Europe.

Islam . . . diffused well beyond its hearth . . . through relocation diffusion of missionaries to portions of sub-Saharan Africa and Southeast Asia. Spatially isolated from the Islamic core region . . . Indonesia . . . is predominantly Muslim, because Arab traders brought the religion there in the thirteenth century.

(199)
Diffusion of Buddhism. Buddhism did not diffuse rapidly from its point of origin in northeastern India. Most responsible for the spread of Buddhism was Asoka, emperor of the Magadhan Empire from about 273 to 232 B.C. About 257 B.C., at the height of the Magadhan Empire's power, Asoka became a Buddhist and thereafter attempted to put into practice Buddha's social principles. In the first century A.D., merchants along the trading routes from northeastern India introduced Buddhism to China. Chinese rulers allowed their people to become Buddhist monks during the fourth century A.D. Buddhism further diffused from China to Korea in the fourth century and from Korea to Japan two centuries later. During the same era, Buddhism lost its original base of support in India.

Diffusion of Other Universalizing Religions. The Bahá'í religion diffused to other regions in the late nineteenth and early twentieth centuries, . . . (then) spread rapidly during the late twentieth century, when a temple was constructed in every continent. Sikhism remained relatively clustered in the Punjab, where the religion originated. In 1802 . . . they created an independent state in the Punjab. But when the British government created the independent states of India and Pakistan in 1947, it divided the Punjab between the two instead of giving the Sikhs a separate country.

Lack of Diffusion of Ethnic Religions. Most ethnic religions have limited, if any, diffusion. These religions lack missionaries. Diffusion of universalizing religions, especially Christianity and Islam, typically comes at the expense of ethnic religions.

Mingling of Ethnic and Universalizing Religions. Universalizing religions may supplant ethnic religions or mingle with them. Equatorial Guinea, a former Spanish colony, is mostly Roman Catholic, whereas Namibia, a former German colony, is heavily Lutheran. Elsewhere, traditional African religious ideas and practices have been merged with Christianity. In East Asia, Buddhism is the universalizing religion that has most mingled with ethnic religions, such as Shintoism in Japan. The current situation in Japan offers a strong caution to anyone attempting to document the number of adherents of any religion. About 90 percent of Japanese say they are Shintos and about 75 percent say they are Buddhists. Ethnic religions can diffuse if adherents migrate to new locations for economic reasons and are not forced to adopt a strongly entrenched universalizing religion. The religious diversity of Mauritius is a function of the country's history of immigration. Mauritius was uninhabited until 1638, so it had no traditional ethnic religion. Hinduism on Mauritius traces back to the Indian immigrants, Islam to the African immigrants, and Christianity to the European immigrants.

(201)
Judaism, an Exception. Only since the creation of the state of Israel in 1948 has a significant percentage of the world's Jews lived in their Eastern Mediterranean homeland. The Romans forced the Jewish diaspora, (from the Greek word for dispersion) after crushing an attempt by the Jews to rebel against Roman rule. Jews lived among other nationalities, retaining separate religious practices

but adopting other cultural characteristics of the host country, such as language. Other nationalities often persecuted the Jews living in their midst. Historically, the Jews of many European countries were forced to live in a **ghetto**, . . . a city neighborhood set up by law to be inhabited only by Jews. During World War II . . . the Nazis systematically rounded up . . . European Jews . . . and exterminated them. Many of the survivors migrated to Israel. Today about 10 percent of the world's 14 million Jews live in Europe, compared to 90 percent a century ago.

Holy Places

Religions may elevate particular places to a holy position. (For) an ethnic religion . . . holy places derive from the distinctive physical environment of its hearth, such as mountains, rivers, or rock formations. A universalizing religion endows with holiness cities and other places associated with the founder's life. Making a **pilgrimage** to these holy places . . . is incorporated into the rituals of some universalizing and ethnic religions.

Holy Places in Universalizing Religions. Buddhism and Islam are the universalizing religions that place the most emphasis on identifying shrines.

(202)
Buddhist Shrines. Eight places are holy to Buddhists because they were the locations of important events in Buddha's life. Because Buddha reached perfect enlightenment while sitting under a bo tree, that tree has become a holy object as well. To honor Buddha, the bo tree has been diffused to other Buddhist countries, such as China and Japan.

Holy Places in Islam. The holiest city for Muslims is Makkah (Mecca), the birthplace of Muhammad.

(203)
The second most holy geographic location in Islam is Madinah (Medina). Muhammad's tomb is at Madinah, inside Islam's second mosque. Every healthy Muslim who has adequate financial resources is expected to undertake a pilgrimage, called a *hajj*, to Makkah (Mecca).

Holy Places in Sikhism. Sikhism's most holy structure, the Darbar Sahib, or Golden Temple, was built at Amritsar, . . . during the seventh century. Militant Sikhs used the Golden Temple . . . as a base for launching attacks in support of greater autonomy . . . during the 1980s.

Holy Places in Ethnic Religions. Ethnic religions are . . . closely tied to the physical geography of a particular place. Pilgrimages are undertaken to view these physical features.

Holy Places in Hinduism. The natural features most likely to rank among the holiest shrines in India are riverbanks or coastlines. Hindus consider a pilgrimage, known as a tirtha, to be an act of purification. Hindus believe that they achieve purification by bathing in holy rivers. The Ganges is the holiest river in India, because it is supposed to spring forth from the hair of Siva. Recent improvements in transportation have increased the accessibility of shrines.

(204)
Cosmogony in Ethnic Religions. Ethnic religions differ from universalizing religions in their understanding of relationships between human beings and nature. These differences derive from distinctive concepts of **cosmogony**, which is a set of religious beliefs concerning the origin of the universe. For example, Chinese ethnic religions, such as Confucianism and Daoism, believe that the universe is made up of two forces, yin and yang, which exist in everything.

The universalizing religions that originated in Southwest Asia, notably Christianity and Islam, consider that God created the universe, including Earth's physical environment and human beings. A religious person can serve God by cultivating the land, draining wetlands, clearing forests, building new settlements, and otherwise making productive use of natural features that God created. In the name of God, some people have sought mastery over nature, not merely independence from it. Large-scale development of remaining wilderness is advocated by some religious people as a way to serve God.

Christians are more likely to consider . . . natural disasters to be preventable and may take steps to overcome the problem by modifying the environment. However, some Christians regard natural disasters as punishment for human sins. Ethnic religions do not attempt to transform the environment to the same extent. Environmental hazards may be accepted as normal and unavoidable.

(205)
The Calendar
Universalizing and ethnic religions have different approaches to the calendar. An ethnic religion typically has . . . holidays based on the distinctive physical geography of the homeland. In universalizing religions, major holidays relate to events in the life of the founder rather than to the changing seasons of one particular place.

The Calendar in Ethnic Religions. A prominent feature of ethnic religions is celebration of the seasons. Rituals are performed to pray for favorable environmental conditions or to give thanks for past success.

The Jewish Calendar. Judaism is classified as an ethnic, . . . religion in part because its major holidays are based on events in the agricultural calendar of the religion's homeland in present-day Israel. The reinterpretation of natural holidays in the light of historical events has been especially important for Jews in the United States, Western Europe, and other regions who are unfamiliar with the agricultural calendar of the Middle East. Israel . . . uses a lunar rather than a solar calendar. The appearance of the new Moon marks the new month in Judaism and Islam and is a holiday for both religions. The lunar month is only about 29 days long, so a lunar year of about 350 days quickly becomes out of step with the agricultural seasons. The Jewish calendar solves the problem by adding an extra month 7 out of every 19 years.

(206)
The Solstice. The **solstice** has special significance in some ethnic religions. A major holiday in some pagan religions is the winter solstice, . . . the shortest day and longest night of the year. Stonehenge . . . is a prominent remnant of a pagan structure apparently aligned so the Sun rises between two stones on the solstice.

The Calendar in Universalizing Religions. The principal purpose of the holidays in universalizing religions is to commemorate events in the founder's life. Christians . . . associate their holidays with seasonal variations, . . . but climate and the agricultural cycle are not central to the liturgy and rituals.

Islamic and Bahá'í calendars. Islam, like Judaism, uses a lunar calendar. Islam as a universalizing religion retains a strict lunar calendar. As a result of using a lunar calendar, Muslim holidays arrive in different seasons from generation to generation. The Bahá'ís use a calendar . . . in which the year is divided into 19 months of 19 days each, with the addition of four intercalary days (five in leap years). The year begins on the first day of spring.

(207)
Christian, Buddhist, and Sikh Holidays. Christians commemorate the resurrection of Jesus on Easter, observed on the first Sunday after the first full Moon following the spring equinox in late March. But not all Christians observe Easter on the same day, because . . . Eastern Orthodox churches use the Julian calendar. Christians may relate Easter to the agricultural cycle, but that relationship differs with where they live. Northern Europeans and North Americans associate Christmas, the birthday of Jesus, with winter conditions. But for Christians in the Southern Hemisphere, December 25 is the height of the summer, with warm days and abundant sunlight. All Buddhists celebrate as major holidays Buddha's birth, Enlightenment, and death. However, Buddhists do not all observe

them on the same days. The major holidays in Sikhism are the births and deaths of the religion's 10 gurus. Commemorating historical events distinguishes Sikhism as a universalizing religion, in contrast to India's ethnic religion, Hinduism, which glorifies the physical geography of India.

Key Issue 3. Why Do Religions Organize Space in Distinctive Patterns?
- **Places of worship**
- **Sacred space**
- **Administration of space**

Geographers study the major impact on the landscape made by all religions, regardless of whether they are universalizing or ethnic. The distribution of religious elements on the landscape reflects the importance of religion in people's values.

Places of Worship
Church, basilica, mosque, temple, pagoda, and synagogue are familiar names that identify places of worship in various religions. Some religions require a relatively large number of elaborate structures, whereas others have more modest needs.

Christian Churches
The Christian landscape is dominated by a high density of churches. The word *church* derives from a Greek term meaning *lord*, *master*, and *power*. *Church* also refers to a gathering of believers, as well as the building where the gathering occurs. The church building plays a more critical role in Christianity than in other religions, in part because the structure is an expression of religious principles, an environment in the image of God . . . (and) because attendance at a collective service of worship is considered extremely important. The prominence of churches on the landscape also stems from their style of construction and location.

(208)
Church Architecture. Early churches were modeled after Roman buildings for public assembly, known as *basilicas*. Churches built during the Gothic period, between the twelfth and fourteenth centuries, had a floor plan in the form of the cross. Since Christianity split into many denominations, no single style of church construction has dominated.

Places of Worship in Other Religions. Unlike Christianity, other major religions do not consider their important buildings a sanctified place of worship.

Muslim Mosques. In contrast to a church, however, a *mosque* is not viewed as a sanctified place but rather as a location for the community to gather together for worship. The mosque is organized around a central courtyard . . . although it may be enclosed in harsher climates. A distinctive feature of the mosque is the minaret, a tower where a man known as a muzzan summons people to worship.

Hindu Temples. In Asian ethnic and universalizing religions . . . important religious functions are . . . likely to take place at home within the family. The Hindu temple serves as a home to one or more gods, although a particular god may have more than one temple. Because congregational worship is not part of Hinduism, the temple does not need a large closed interior space filled with seats. The site of the temple . . . may also contain . . . a pool for ritual baths.

Buddhist and Shintoist Pagodas. The pagoda is a prominent and visually attractive element of the Buddhist and Shintoist landscapes. Pagodas contain relics that Buddhists believe to be a portion of Buddha's body or clothing. Pagodas are not designed for congregational worship.

(209)
Bahá'í Houses of Worship. Bahá'ís built seven Houses of Worship . . . dispersed to different continents to dramatize Bahá'í as a universalizing religion, . . . open to adherents of all religions. Services include reciting the scriptures of various religions.

Sacred Space

The impact of religion is clearly seen . . . at several scales. How each religion distributes its elements on the landscape depends on its beliefs.

Disposing of the Dead. A prominent example of religiously inspired arrangement of land at a smaller scale is burial practices.

Burial. Christians, Muslims, and Jews usually bury their dead in a specially designated area called a *cemetery*. After Christianity became legal, Christians buried their dead in the yard around the church. Public health and sanitation considerations in the nineteenth century led to public management of many cemeteries. The remains of the dead are customarily aligned in some traditional direction. In congested urban areas, Christians and Muslims have traditionally used cemeteries as public open space. Traditional burial practices in China . . . have removed as much as 10 percent of the land from productive agriculture.

Other Methods of Disposing of Bodies. Not all faiths bury their dead. Hindus generally practice cremation rather than burial. Cremation was the principal form of disposing of bodies in Europe before Christianity. Motivation for cremation may have originated from unwillingness on the part of nomads to leave their dead behind. Cremation could also free the soul from the body. To strip away unclean portions of the body, Parsis (Zoroastrians) expose the dead to scavenging birds and animals. Tibetan Buddhists also practice exposure for some dead, with cremation reserved for the most exalted priests. Disposal of bodies at sea is used in some parts of Micronesia, but the practice is much less common than in the past.

(210)

Religious Settlements. Buildings for worship and burial places are smaller-scale manifestations of religion on the landscape, but there are larger-scale examples: entire settlements. A utopian settlement is an ideal community built around a religious way of life. By 1858 some 130 different utopian settlements had begun in the United States.

Most utopian communities declined in importance or disappeared altogether. Although most colonial settlements were not planned primarily for religious purposes, religious principles affected many of the designs. New England settlers placed the church at the most prominent location in the center of the settlement.

(211)

Religious Place Names. Roman Catholic immigrants frequently have given religious place names, or toponyms, to their settlements in the New World, particularly in Québec and the U.S. Southwest.

Administration of Space

Followers of a universalizing religion must be connected so as to assure communication and consistency of doctrine. Ethnic religions tend not to have organized, central authorities.

Hierarchical Religions. A hierarchical religion has a well-defined geographic structure and organizes territory into local administrative units.

Roman Catholic Hierarchy. The Roman Catholic Church has organized much of Earth's inhabited land into an administrative structure, ultimately accountable to the Pope in Rome. Reporting to the Pope are *archbishops*. Each archbishop heads a *province*, which is a group of several *dioceses*. Reporting to each archbishop are *bishops*. Each bishop administers a diocese, of which there are several thousand. A diocese in turn is spatially divided into parishes, each headed by a priest.

(212)

Latter-Day Saints. Latter-Day Saints (Mormons) exercise strong organization of the landscape. The highest authority in the Church . . . frequently redraws ward and stake boundaries in rapidly growing areas to reflect the ideal population standards.

Locally Autonomous Religions. Some universalizing religions are highly **autonomous religions**, or self-sufficient, and interaction among communities is confined to little more than loose cooperation and shared ideas. Islam and some Protestant denominations are good examples.

Local Autonomy in Islam. Islam has neither a religious hierarchy nor a formal territorial organization. Strong unity within the Islamic world is maintained by a relatively high degree of communication and migration, such as the pilgrimage to Makkah. In addition, uniformity is fostered by Islamic doctrine, which offers more explicit commands than other religions.

Protestant Denominations. Protestant Christian denominations vary in geographic structure from extremely autonomous to somewhat hierarchical. Extremely autonomous denominations such as Baptists and United Church of Christ are organized into self-governing congregations. Presbyterian churches represent an intermediate degree of autonomy. The Episcopalian, Lutheran, and most Methodist churches have hierarchical structures, somewhat comparable to the Roman Catholic Church.

Ethnic Religions. Judaism and Hinduism also have no centralized structure of religious control.

Key Issue 4. Why Do Territorial Conflicts Arise among Religious Groups?
- **Religion vs. government policies**
- **Religion vs. religion**

The twentieth century was a century of global conflict. The threat of global conflict has receded in the twenty-first century, but local conflicts have increased in areas of cultural diversity, as will be discussed in Chapters 7 and 8. The element of cultural diversity that has led to conflict in many localities is religion. In a world increasingly dominated by a global culture and economy, religious fundamentalism is one of the most important ways that a group maintains a distinctive cultural identity.

(213)
Religion vs. Government Policies. The role of religion in organizing Earth's surface has diminished in some societies, owing to political and economic change. Yet in recent years religious principles have become increasingly important in the political organization of countries, especially where a branch of Christianity or Islam is the prevailing religion.

Religion vs. Social Change. Participation in the global economy and culture can expose local residents to values and beliefs originating in more developed countries . . . . North Americans and Western Europeans may not view economic development as incompatible with religious values, but many religious adherents in less developed countries do, especially where Christianity is not the predominant religion.

Hinduism vs. Social Equality. Hinduism has been strongly challenged since the 1800s, when British colonial administrators introduced their social and moral concepts to India. The most vulnerable aspect of the Hindu religion was its rigid **caste** system. British administrators and Christian missionaries pointed out the shortcomings of the caste system, such as neglect of the untouchables' health and economic problems. The Indian government legally abolished the untouchable caste, and the people formerly in that caste now have equal rights with other Indians.

(214)
Religion vs. Communism. Organized religion was challenged in the twentieth century by the rise of communism in Eastern Europe and Asia.

Eastern Orthodox Christianity and Islam vs. the Soviet Union. In 1721 Czar Peter the Great made the Russian Orthodox Church a part of the Russian government. Following the 1917 Bolshevik revolution, which overthrew the czar, the Communist government of the Soviet Union pursued antireligious programs. People's religious beliefs could not be destroyed overnight, but the role of organized religion in Soviet life was reduced.

All church buildings and property were nationalized and could be used only with local government permission. With religious organizations prevented from conducting social and cultural work, religion dwindled in daily life. The end of Communist rule in the late twentieth century brought a religious revival in Eastern Europe, especially where Roman Catholicism is the most prevalent branch. Property confiscated by the Communist governments reverted to Church ownership, and attendance at church services increased. Central Asian countries that were former parts of the Soviet Union . . . are struggling to determine the extent to which laws should be rewritten to conform to Islamic custom rather than to the secular tradition inherited from the Soviet Union.

Buddhism vs. Southeast Asian Countries. In Southeast Asia, Buddhists were hurt by the long Vietnam War. Neither antagonist was particularly sympathetic to Buddhists. The current Communist governments in Southeast Asia have discouraged religious activities and permitted monuments to decay. These countries do not have the funds necessary to restore the structures.

Religion vs. Religion

Conflicts are most likely to occur (at) . . . a boundary between two religious groups. Two long-standing conflicts involving religious groups are in the Middle East and Northern Ireland.

Religious Wars in the Middle East. Jews, Christians, and Muslims have fought for 2,000 years. All three religions have especially strong attachments to the city of Jerusalem. As an ethnic religion, Judaism makes a special claim to the territory it calls the Promised Land. The religion's customs and rituals acquired meaning from the agricultural life of the ancient Hebrew tribe. Jerusalem is especially holy to Jews because it was the location of the Temple, their center of worship in ancient times. Christians consider Palestine the Holy Land and Jerusalem the Holy City because the major events in Jesus' life, death, and Resurrection were concentrated there. Muslims regard Jerusalem as their third holiest city, after Makkah and Madinah. The most important Muslim structure in Jerusalem is the mosque at the Dome of the Rock, built in 691. The rock is thought to be the place from which Muhammad ascended to heaven.

Crusades between Christians and Muslims. In the seventh century, Muslims . . . captured most of the Middle East . . . and converted most of the people from Christianity to Islam. The Arab army moved west across North Africa and invaded Europe at Gibraltar in 710 . . . (and) crossed the Pyrenees Mountains a few years later. Its initial advance in Europe was halted by the Franks . . . led by Charles Martel, at Poitiers, France, in 732. The Arab army . . . continued to control portions of present-day Spain until 1492

(216)
To the east, the Arab army captured Eastern Orthodox Christianity's most important city, Constantinople (present-day Istanbul in Turkey), in 1453 and advanced a few years later into Southeast Europe, as far north as present-day Bosnia and Herzegovina. The current civil war in that country is a legacy of the fifteenth-century Muslim invasion. To recapture the Holy Land from its Muslim conquerors, European Christians launched a series of military campaigns, known as Crusades.

Jews vs. Muslims in Palestine. The Muslim Ottoman Empire controlled Palestine . . . between 1516 and 1917. Great Britain took over Palestine under a mandate from the League of Nations. The British allowed some Jews to return to Palestine, but immigration was restricted again during the 1930s in response to intense pressure by Arabs in the region. As violence initiated by both Jewish and Muslim settlers escalated after World War II, the British announced their intention to withdraw from Palestine. The United Nations voted to partition Palestine into two independent states. Jerusalem was to be an international city, open to all religions, and run by the United Nations. When the British withdrew in 1948, Jews declared an independent state of Israel within the boundaries prescribed by the U.N. resolution. The next day its neighboring Arab Muslim states declared war. The combatants signed an armistice in 1949 that divided control of Jerusalem. Israel won three more wars with its neighbors, in 1956, 1967, and 1973. During the 1967 Six-Day War, Israel captured the entire city of Jerusalem and removed the barriers that had prevented Jews from visiting and living in the Old City of Jerusalem. The ultimate obstacle to comprehensive peace in the Middle East is the status of Jerusalem.

(217)

Conflict over the Holy Land: Palestinian Perspective. After the 1973 war, Egypt and Jordan signed peace treaties with Israel, and Syria stopped actively plotting an attack on Israel. Despite the movement toward peace among the neighboring nationalities in the Middle East, unrest persists because of the emergence of a new nationality in the late 1960s, known as the Palestinians. To complicate the situation, five groups of people consider themselves Palestinians. After capturing the West Bank from Jordan in 1967, Israel permitted Jewish settlers to construct more than 100 settlements in the territory. Although Jewish settlers comprise only about 7 percent of the West Bank population, Palestinians see their presence as a reflection of Israel's reluctance to grant independence to the occupied territory.

Israel has turned over control of the Gaza and much of the West Bank to the Palestine Liberation Organization (PLO). Some Palestinians . . . are willing to settle for all of the territory taken by Israel in the 1967 War, including the Old City of Jerusalem, while others want to continue fighting Israel for the entire territory between the Jordan River and the Mediterranean Sea.

Conflict over the Holy Land: Israeli Perspective. Israel sees itself as a very small country . . . with a Jewish majority, surrounded by a region of hostile Muslim Arabs. Israel considers two elements of the local landscape especially meaningful. First, the country's major population centers are quite close to international borders, making them vulnerable to surprise attack. The second geographical problem from Israel's perspective derives from local landforms.

(218)

The partition of Palestine in 1947. . . allocated most of the coastal plain to Israel, while Jordan took most of the hills between the coastal plain and the Jordan River, . . . called the West Bank (of the Jordan River). Farther north, Israel's territory extended eastward to the Jordan River valley, but Syria controlled the highlands east of the valley, known as the Golan Heights. Between 1948 and 1967 Jordan and Syria used the hills as staging areas to attack Israeli settlements on the adjacent coastal plain and in the Jordan River valley. During the 1967 War, Israel captured these highlands to stop attacks on the lowland population concentrations. Israeli Jews are divided between those who wish to retain some of the occupied territories and those who wish to make compromises with the Palestinians. Peace   . . . will be difficult to achieve because Israelis have no intention of giving up control of the Old City of Jerusalem, and Palestinians have no intention of giving up their claim to it.

(220)

Religious Wars in Ireland. The most troublesome religious boundary in Western Europe lies on the island of Eire (Ireland). The Republic of Ireland, which occupies five-sixths of the island, is 92 percent Roman Catholic, but the island's northern one-sixth, which is part of the United Kingdom rather than Ireland, is about 58 percent Protestant and 42 percent Roman Catholic.

Ireland became a self-governing dominion within the British Empire in 1921. Complete independence was declared in 1937, and a republic was created in 1949. When most of Ireland became independent, a majority in six northern counties voted to remain in the United Kingdom. Demonstrations by Roman Catholics protesting discrimination began in 1968. Since then, more than 3,000 have been killed in Northern Ireland—both Protestants and Roman Catholics. A small number of Roman Catholics in both Northern Ireland and the Republic of Ireland joined the Irish Republican Army

(IRA), a militant organization dedicated to achieving Irish national unity by whatever means available, including violence. Similarly, a scattering of Protestants created extremist organizations to fight the IRA, including the Ulster Defense Force (UDF). As long as most Protestants are firmly committed to remaining in the United Kingdom and most Roman Catholics are equally committed to union with the Republic of Ireland, peaceful settlement appears difficult.

Key Terms

Animism (p.194)
Autonomous religion (p.212)
Branch (p.187)
Caste (p.213)
Cosmogony (p.204)
Denomination (p.187)
Diocese (p.211)
Ethnic religion (p.212)
Fundamentalism (p.212)
Ghetto (p.201)

Hierarchical religion (p.211)
Missionary (p.197)
Monotheism (p.194)
Pagan (p.197)
Pilgrimage (p.201)
Polytheism (p.194)
Sect (p.187)
Solstice (p.206)
Universalizing religion (p.187)

Chapter 7. Ethnicity

Ethnicity is a source of pride to people, a link to the experiences of ancestors and to cultural traditions. The ethnic group to which one belongs has important measurable differences. Ethnicity also matters in places with a history of discrimination by one ethnic group against another.

The significance of ethnic diversity is controversial in the United States:
- To what extent does discrimination persist against minority ethnicities?
- Should preferences be given to minority ethnicities to correct past patterns of discrimination?
- To what extent should the distinct cultural identity of ethnicities be encouraged or protected?

Key Issues
1. Where are ethnicities distributed?
2. Why have ethnicities been transformed into nationalities?
3. Why do ethnicities clash?
4. What is ethnic cleansing?

(227)
Ethnicity is identity with a group of people who share the cultural traditions of a particular homeland or hearth. Ethnicity comes from the Greek word *ethnikos*, which means national. Geographers are interested in *where* ethnicities are distributed across *space*, like other elements of culture. Like other cultural elements, ethnic identity derives from the interplay of *connections* with other groups and isolation from them. Ethnicity is an especially important cultural element of *local diversity* because our ethnic identity is immutable. The study of ethnicity lacks the tension in *scale* between preservation of local diversity and *globalization* observed in other cultural elements. No ethnicity is attempting or even aspiring to achieve global dominance. In the face of globalization . . . ethnicity stands as the strongest bulwark for the preservation of local diversity.

Key Issue 1. Where Are Ethnicities Distributed?
- **Distribution of ethnicities in the United States**
- **Differentiating ethnicity and race**

This section of the chapter examines the clustering of ethnicities within countries, and the next key issue looks at ethnicities at the national scale.

Distribution of Ethnicities in the United States
The two most numerous ethnicities in the United States are African-Americans, about 13 percent, and Hispanics or Latinos, about 11 percent. In addition, about 4 percent are Asian-American and 1 percent American Indian.

Clustering of Ethnicities
Clustering of ethnicities can occur at two scales, . . . particular regions of the country, and . . . particular neighborhoods within cities.

Regional Concentrations of Ethnicities. African-Americans are clustered in the Southeast, Hispanics in the Southwest, Asian-Americans in the West, and American Indians in the Southwest and Plains states. African-Americans comprise at least one-fourth of the population in Alabama, Georgia, Louisiana, and South Carolina, and more than one-third in Mississippi. At the other extreme, 9 states have fewer than 1 percent African-Americans.

Hispanic or *Hispanic-American* is a term that the U.S. government chose in 1973 . . . because it was an inoffensive label that could be applied to all people from Spanish-speaking countries. Some Americans of Latin-American descent have adopted the term Latino instead.

(228)

Most Hispanics identify with a more specific ethnic or national origin. Within the United States, Hispanics are heavily clustered in the four southwestern states. About 4 percent of the U.S. population is Asian-American. Chinese account for about 25 percent of Asian-Americans, Filipinos 20 percent, and Japanese, Asian Indians, and Vietnamese 12 percent each. The largest concentration of Asian-Americans is in Hawaii. One-half of all Asian-Americans live in California.

American Indians and Alaska Natives make up about 1 percent of the U.S. population. Within the 48 continental United States, American Indians are most numerous in the Southwest and the Plains states.

Concentration of Ethnicities in Cities. About one-fourth of all Americans live in cities, whereas more than half of African-Americans live in cities. The contrast is greater at the state level. For example, African-Americans comprise three-fourths of the population in the city of Detroit and only one-twentieth in the rest of Michigan. The distribution of Hispanics is similar to that of African-Americans in large northern cities.

(229)

In the states with the largest Hispanic populations—California and Texas—the distribution is mixed. The clustering of ethnicities is especially pronounced at the scale of neighborhoods within cities. During the twentieth century the children and grandchildren of European immigrants moved out of most of the original inner-city neighborhoods. For descendants of European immigrants, ethnic identity is more likely to be retained through religion, food, and other cultural traditions rather than through location of residence. Ethnic concentrations in U.S. cities increasingly consist of African-Americans who migrate from the South, or immigrants from Latin America and Asia. In Los Angeles, which contains large percentages of African-Americans, Hispanics, and Asian-Americans, the major ethnic groups are clustered in different areas.

(230)
African-American Migration Patterns
Three major migration flows have shaped (African-American) distribution within the United States: immigration from Africa . . . in the eighteenth century; immigration . . . to northern cities during the first half of the twentieth century; (and) immigration from inner-city ghettos to other urban neighborhoods in the second half of the twentieth century.

Forced Migration from Africa. The first Africans brought to the American colonies as slaves . . . (arrived in) 1619. During the eighteenth century the British shipped about 400,000 Africans to the 13 colonies. In 1808 the U.S. banned bringing in . . . slaves, but an estimated 250,000 were illegally imported during the next half century. Slavery was replaced in Europe by a feudal system, in which laborers . . . were bound to the land and not free to migrate elsewhere. Although slavery was rare in Europe, Europeans were responsible for diffusing the practice to the Western Hemisphere, . . . a response to a shortage of labor in the sparsely inhabited Americas.

The forced migration began when people living along the east and west coasts of Africa, taking advantage of their superior weapons, captured members of other groups living farther inland and sold the captives to Europeans. Fewer than 5 percent of the slaves ended up in the United States. At the height of the eighteenth-century slave demand, a number of European countries adopted the triangular slave trade. Some ships added another step, making a rectangular trading pattern, in which molasses was carried from the Caribbean to the North American colonies, and rum from the colonies to Europe.

(231)
In the 13 colonies that later formed the United States, most of the large plantations in need of labor were located in the South, primarily those growing cotton as well as tobacco. Attitudes toward slavery dominated U.S. politics during the nineteenth century. The Civil War (1861–1865) was fought to prevent 11 pro-slavery southern states from seceding from the Union. Freed as slaves, most African-Americans remained in the rural South during the late nineteenth century working as sharecroppers. A **sharecropper** works fields rented from a landowner and pays the rent by turning over to the

landowner a share of the crops. The sharecropper system burdened poor African-Americans with high interest rates and heavy debts. Instead of growing food that they could eat, sharecroppers were forced by landowners to plant extensive areas of crops such as cotton that could be sold for cash.

Immigration to the North. Sharecropping declined in the early twentieth century as . . . farm machinery and decline in . . . cotton reduced demand for labor. At the same time sharecroppers were being pushed off the farms, they were being pulled to the prospect of jobs in the booming industrial cities of the North. African-Americans migrated out of the South along several clearly defined channels . . . along the major two-lane long-distance U.S. roads that had been paved and signposted in the early decades of the twentieth century. Southern African-Americans migrated north and west in two main waves, the first in the 1910s and 1920s before and after World War I and the second in the 1940s and 1950s before and after World War II.

(232)

Expansion of the Ghetto. When they reached the big cities, African-American immigrants clustered in the one or two neighborhoods where the small numbers who had arrived in the nineteenth century were already living. These areas became known as ghettos, after the term for neighborhoods in which Jews were forced to live in the Middle Ages (see Chapter 6). African-Americans moved from the tight ghettos into immediately adjacent neighborhoods during the 1950s and 1960s.

Differentiating Ethnicity and Race

Ethnicity is distinct from **race**, which is identity with a group of people who share a biological ancestor. Race comes from a middle-French word for *generation*. Race and ethnicity are often confused. In the United States, consider three prominent ethnic groups—Asian-Americans, African-Americans, and Hispanic-Americans. Asian as a race and Asian-American as an ethnicity encompass basically the same group. African-American and black are different groups. Some American blacks . . . trace their cultural heritage to regions other than Africa, including Latin America, Asia, or Pacific islands. Hispanic or Latino is not considered a race.

(231)

The traits that characterize race are those that can be transmitted genetically from parents to children: lactose intolerance, for example. Biological features of all humans . . . were once thought to be scientifically classifiable into a handful of world races. Biological features are so highly variable among members of a race that any prejudged classification is meaningless. The degree of isolation needed to keep biological features distinct genetically vanished when the first human crossed a river or climbed a hill. At worst, biological classification by race is the basis for **racism**, which is the belief that race is the primary determinant of human traits and capacities and that racial differences produce an inherent superiority of a particular race.

Ethnicity is important to geographers because its characteristics derive from the distinctive features of particular places on Earth. In contrast, contemporary geographers reject the entire biological basis of classifying humans . . . because these features are not rooted in specific places. One feature of race does matter to geographers—the color of skin. The distribution of persons of color matters . . . because it is the most fundamental basis by which people in many societies sort out where they reside, attend school, recreate, and perform many other activities of daily life. The term *African-American* identifies a group with an extensive cultural tradition, whereas the term *black* in principle denotes nothing more than a dark skin.

Race in the United States

Every 10 years the U.S. Bureau of the Census asks people to classify themselves according to races with which they most closely identify. The 2000 census permitted people to check more than 1 of 14 categories listed. A distinctive feature of race relations in the United States has been the strong discouragement of spatial interaction—in the past through legal means, today through cultural preferences or discrimination.

(234)
"Separate but Equal" Doctrine. In 1896 the U.S. Supreme Court upheld a Louisiana law that required black and white passengers to ride in separate railway cars, in Plessy v. Ferguson. Once the Supreme Court permitted "separate but equal" treatment of the races, southern states enacted a comprehensive set of laws to segregate blacks from whites as much as possible. Throughout the country, not just in the South, house deeds contained restrictive covenants that prevented the owners from selling to blacks, as well as to Roman Catholics or Jews in some places.

"White Flight." Segregation laws were eliminated during the 1950s and 1960s. The landmark Supreme Court decision, Brown v. Board of Education of Topeka, Kansas, in 1954, found that separate schools for blacks and whites was unconstitutional. A year later the Supreme Court further ruled that schools had to be desegregated "with all deliberate speed." Rather than integrate, whites fled. The expansion of the black ghettos in American cities was made possible by "white flight." Detroit provides a clear example. In the late 1960s the National Advisory Commission on Civil Disorders . . . concluded that U.S. cities were divided into two separate and unequal societies. Three decades later . . . segregation and inequality persist.

(235)
Division by Race in South Africa
Discrimination by race reached its peak in the late twentieth century in South Africa. **Apartheid** was the physical separation of different races into different geographic areas. Although South Africa's apartheid laws were repealed during the 1990s, it will take many years for it to erase the impact of past policies.
Apartheid System. In South Africa, under apartheid, a newborn baby was classified as being one of four races: black, white, colored (mixed white and black), or Asian. According to the most recent census, blacks constitute about 76 percent of South Africa's population, whites 13 percent, colored 9 percent, and Asians 3 percent. Under apartheid, each of the four races had a different legal status in South Africa. The apartheid system was created by descendants of whites who arrived in South Africa from Holland in 1652. They were known either as *Boers*, from the Dutch word for *farmer*, or *Afrikaners*, from the word "Afrikaans," the name of their language, which is a dialect of Dutch. A series of wars between the British and the Boers culminated in a British victory in 1902, and South Africa became part of the British Empire.

(236)
British descendants continued to control South Africa's government until 1948, when the Afrikaner-dominated Nationalist Party won elections. Colonial rule was being replaced in the rest of Africa by a collection of independent states run by the local black population. The Nationalist Party created the apartheid laws in the next few years to perpetuate white dominance of the country. To assure further geographic isolation of different races, the South African government designated 10 so-called *homelands* for blacks. If the government policy had been fully implemented, the 10 black homelands together would have contained approximately 44 percent of South Africa's population on only 13 percent of the land.

Dismantling of Apartheid. In 1991 the white-dominated government of South Africa repealed the apartheid laws, including restrictions on property ownership and classification of people at birth by race. The African National Congress was legalized, and its leader, Nelson Mandela, was released from jail after more than 27 years. When all South Africans were permitted to vote in national elections for the first time, in April 1994, Mandela was overwhelmingly elected the country's first black president. Whites were guaranteed representation in the government during a five-year transition period, until 1999. Now that South Africa's apartheid laws have been dismantled and the country is governed by its black majority, other countries have reestablished economic and cultural ties. However, the legacy of apartheid will linger for many years. Average income among white South Africans is about 10 times higher than for blacks.

(237)

Key Issue 2. Why Have Ethnicities Been Transformed into Nationalities?
- **Rise of nationalities**
- **Nationalities in former colonies**
- **Revival of ethnic identity**

Ethnicity is distinct from race and nationality, two other terms commonly used to describe a group of people with shared traits. **Nationality** is identity with a group of people who share legal attachment and personal allegiance to a particular country. It comes from the Latin word *nasci*, which means *to have been born*. In principle, the cultural values shared with others of the same ethnicity derive from religion, language, and material culture, whereas those shared with others of the same nationality derive from voting, obtaining a passport, and performing civic duties. In the United States, nationality is generally kept reasonably distinct from ethnicity and race in common usage.

In Canada the Québécois are clearly distinct from other Canadians in language, religion, and other cultural traditions. But do the Québécois form a distinct ethnicity within the Canadian nationality or a second nationality separate altogether from Anglo-Canadian? The distinction is critical. Outside North America, distinctions between ethnicity and nationality are even muddier. Confusion between ethnicity and nationality can lead to violent conflicts.

Rise of Nationalities
Descendants of nineteenth-century immigrants to the United States from central and Eastern Europe identify themselves today by ethnicity rather than by nationality. These ethnicities lived in Europe as subjects of the Austrian emperor, Russian czar, or Prussian kaiser. U.S. immigration officials recorded the nationality of immigrants. But immigrants considered ethnicity more important than nationality, and that is what they have preserved through distinctive social customs. The United States forged a nation in the late eighteenth century out of a collection of ethnic groups. To be an American meant believing in the "unalienable rights" of "life, liberty, and the pursuit of happiness."

Nation-States
The concept that ethnicities have the right to govern themselves is known as **self-determination**. During the nineteenth and twentieth centuries, political leaders have generally supported the right of self-determination . . . and have attempted to organize Earth's surface into a collection of **nation-states** . . . whose territory corresponds to . . . a particular ethnicity. Yet despite continuing attempts, . . . the territory of a state rarely corresponds precisely to the territory occupied by an ethnicity.

(238)

Nation-States in Europe. Ethnicities were transformed into nationalities throughout Europe during the nineteenth century. Most of Western Europe was made up of nation-states by 1900. Following their defeat in World War I, the Austro-Hungarian and Ottoman empires were dismantled, and many European boundaries were redrawn according to the principle of nation-states. During the 1930s, German National Socialists (Nazis) claimed that all German-speaking parts of Europe constituted one nationality and should be unified into one state. Other European powers did not attempt to stop the Germans from taking over Austria and the German-speaking portion of Czechoslovakia. Not until the Germans invaded Poland (clearly not a German-speaking country) in 1939 did England and France try to stop them.

Denmark: There Are No Perfect Nation-States. Denmark is a fairly good example of a European nation-state. The territory occupied by the Danish ethnicity closely corresponds to the state of Denmark. But even Denmark is not a perfect example of a nation-state. The country's . . . southern boundary with Germany does not divide Danish and German nationalities precisely. Denmark controls two territories in the Atlantic Ocean that do not share Danish cultural characteristics—the Faeroe Islands and Greenland. In 1979 Greenlanders received more authority . . . to control their own domestic affairs. One decision was to change all place names in Greenland from Danish to the local Inuit language.

Nationalism

A nationality, once established, must hold the loyalty of its citizens to survive. **Nationalism** typically promotes a sense of national consciousness that exalts one nation above all others. For many states, mass media are the most effective means of fostering nationalism. Consequently, only a few states permit mass media to operate without government interference.

(239)

Nationalism can have a negative impact. The sense of unity within a nation-state is sometimes achieved through the creation of negative images of other nation-states. Nationalism is an important example of a centripetal force, which is an attitude that tends to unify people and enhance support for a state. (The word centripetal means "directed toward the center." It is the opposite of centrifugal, which means to spread out from the center.)

Multinational States

In some **multi-ethnic states**, ethnicities all contribute cultural features to the formation of a single nationality. Belgium is divided among the Dutch-speaking Flemish and the French-speaking Walloons. Both groups consider themselves belonging to the Belgian nationality. Other multi-ethnic states, known as **multinational states**, contain two ethnic groups with traditions of self-determination that agree to coexist peacefully by recognizing each other as distinct nationalities. One example of a multinational state is the United Kingdom, which contains four main nationalities—England, Scotland, Wales, and Northern Ireland.

Today the four nationalities hold little independent political power, although Scotland and Wales now have separately elected governments. The main element of distinct national identity comes from sports. Given the history of English conquest, the other nationalities typically root against England when it is playing teams from other countries. Ethnicities do not always find ways to live together peacefully. In some cases, ethnicities compete in civil wars to dominate the national identity. In other cases, problems result from confusion between ethnic identity and national identity.

Former Soviet Union: The Largest Multinational State

The Soviet Union was an especially prominent example of a multinational state until its collapse in the early 1990s.

The 15 republics that once constituted the Soviet Union are now independent countries. When the Soviet Union existed, its 15 republics were based on the 15 largest ethnicities. Less numerous ethnicities were not given the same level of recognition. With the breakup . . . a number of these less numerous ethnicities are now divided among more than one state. The 15 newly independent states consist of five groups, 3 Baltic, 3 European, 5 Central Asian, 3 Caucasus, (and) Russia. Reasonably good examples of nation-states have been carved out of the Baltic, European, and some Central Asian states (but not). . . in any of the small Caucasus states, and Russia is an especially prominent example of a state with major difficulties in keeping all of its ethnicities contented.

(240)

New Baltic Nation-States. Estonia, Latvia, and Lithuania . . . had been independent countries between . . . 1918 and 1940. Of the three Baltic states, Lithuania most closely fits the definition of a nation-state, because 81 percent of its population are ethnic Lithuanians. These three small neighboring Baltic countries have clear cultural differences and distinct historical traditions.

New European Nation-States. To some extent, the former Soviet republics of Belarus, Moldova, and Ukraine now qualify as nation-states. The ethnic distinctions among Belarusians, Ukrainians, and Russians are somewhat blurred. Belarusians and Ukrainians became distinct ethnicities because they were isolated from the main body of Eastern Slavs—the Russians—during the thirteenth and fourteenth centuries. Russians actually constitute two-thirds of the population in the Crimean Peninsula of Ukraine. After Russia and Ukraine became separate countries, a majority of the Crimeans voted to become independent of Ukraine. Control of the Crimean Peninsula was also important to both Russia and Ukraine because one of the Soviet Union's largest fleets was stationed there. The two countries agreed to divide the ships and to jointly maintain the naval base at Sevastopol. Compounding the problem in the Crimea, 166,000 Tatars have migrated there from Central Asia in recent years. The Tatars once lived in the Crimea, but the Soviet leadership . . . deported them to Central Asia. The Tatars prefer to be governed by Ukraine.

The situation is different in Moldova. Moldovans are ethnically indistinguishable from Romanians, and Moldova (then called Moldavia) was part of Romania until the Soviet Union seized it in 1940. In 1992, many Moldovans pushed for reunification with Romania. But it was not to be that simple. The Soviet government increased the size of Moldova by about 10 percent, transferring from Ukraine a sliver of land on the east bank of the Dniester (River). Inhabitants of this area are Ukrainian and Russian. They oppose Moldova's reunification with Romania.

(241)
New Central Asian States. The five states in Central Asia carved out of the former Soviet Union display varying degrees of conformance to the principles of nation-state. Together the five provide an important reminder that multinational states can be more peaceful than nation-states.

In Turkmenistan and Uzbekistan, the leading ethnic group has an overwhelming majority—77 percent Turkmen and 80 percent Uzbek, respectively. Turkmen and Uzbeks are examples of ethnicities split into more than one country, the Turkmen between Turkmenistan and Russia, and Uzbeks among Kyrgyzstan, Tajikistan, and Uzbekistan.

Kyrgyzstan is 52 percent Kyrgyz, 18 percent Russian, and 13 percent Uzbek. The Kyrgyz—also Muslims who speak an Altaic language—resent the Russians for seizing the best farmland.

In principle, Kazakhstan, twice as large as the other four Central Asian countries combined, is a recipe for ethnic conflict. The country is divided almost evenly between Kazakhs and Russians. Kazakhstan has been peaceful, in part because it has a somewhat less depressed economy than its neighbors.

In contrast, Tajikistan—65 percent Tajik, 25 percent Uzbek, and only 3 percent Russian—would appear to be a stable country, but it suffers from a civil war among the Tajik people. The civil war has been between Tajiks who are former Communists and an unusual alliance of Muslim fundamentalists and Western-oriented intellectuals.

Russia: Now the Largest Multinational State
Russia officially recognizes the existence of 39 nationalities, many of which are eager for independence. Russia's ethnicities are clustered in two principal locations. Some are located along borders with neighboring states.

(242)
Other ethnicities are clustered in the center of Russia. Most of these groups were conquered by the Russians in the sixteenth century. Independence movements are flourishing, because Russia is less willing to suppress these movements forcibly than the Soviet Union had once been. Particularly troublesome . . . are the Chechens, a group of Sunni Muslims who speak a Caucasian language and practice distinctive social customs. Chechnya was brought under Russian control in the nineteenth century only after a 50-year fight. When the Soviet Union broke up . . . the Chechens declared their independence. Russia fought hard to prevent Chechnya from gaining independence because it feared that other ethnicities would follow suit. Chechnya was also important to Russia because the region contained deposits of petroleum.

Russians in Other States. Decades of Russian domination has left a deep reservoir of bitterness among other ethnicities once part of the Soviet Union. Russian soldiers have remained stationed in other countries, in part because Russia cannot afford to rehouse them. Other ethnicities fear . . . the Russians are trying to reassert . . . dominance. For their part, Russians claim that they are now subject to discrimination as minorities in countries that were once part of the Soviet Union. Russians living in other countries of the former Soviet Union feel that they cannot migrate to Russia, because they have no jobs, homes, or land awaiting them there.

Turmoil in the Caucasus

The Caucasus region . . . gets its name from the mountains that separate Russia from Azerbaijan and Georgia. The region is home to several ethnicities. Each ethnicity has a long-standing and complex set of grievances against others in the region. Every ethnicity in the Caucasus has the same aspiration: to carve out a sovereign nation-state.

Azeris. Azeris (or Azerbaijanis) trace their roots to Turkish invaders . . . in the eighth and ninth centuries. An 1828 treaty allocated northern Azeri territory to Russia and southern Azeri territory to Persia (now Iran). More than 7 million Azeris now live in Azerbaijan, 90 percent of the country's total population. Another 6 million Azeris are clustered in northwestern Iran. Azeris hold positions of responsibility in Iran's government and economy, but Iran restricts teaching of the Azeri language.

Armenians. More than 3,000 years ago Armenians controlled an independent kingdom in the Caucasus. During the late nineteenth and early twentieth centuries, hundreds of thousands of Armenians were killed in a series of massacres organized by the Turks. Others were forced to migrate to Russia. After World War I the allies created an independent state of Armenia, but it was soon swallowed by its neighbors. Turkey and the Soviet Union . . . divided Armenia. The Soviet portion became . . . an independent country in 1991. More than 90 percent of the population in Armenia are Armenians, making it the most ethnically homogeneous country in the region. Armenians and Azeris . . . have been at war with each other since 1988 over the boundaries between the two nationalities.

(243)

Georgians. The population of Georgia is more diverse than that in Armenia and Azerbaijan. Georgia's cultural diversity has been a source of unrest, especially among the Ossetians and Abkhazians. Abkhazians want an independent state in the northwest, while the Ossetians want to rejoin the south of Georgia to Russia.

Revival of Ethnic Identity

Ethnic identities never really disappeared in Africa, where loyalty to tribe often remained more important than loyalty to the nationality of a new country, perhaps controlled by another ethnicity. Europeans thought that ethnicity had been left behind as an insignificant relic, such as wearing quaint costumes to amuse tourists. But Europeans were wrong.

Ethnicity and Communism

From the end of World War II in 1945 until the early 1990s, attitudes toward communism and economic cooperation were more important political factors in Europe than the nation-state principle. For example, the Communist government of Bulgaria repressed cultural differences by banning the Turkish language and the practice of some Islamic religious rites . . . to remove . . . obstacles to unifying national support for the ideology of communism.

The Communists did not completely suppress ethnicities in Eastern Europe: The administrative structures of the former Soviet Union and two other multi-ethnic Eastern European countries—Czechoslovakia and Yugoslavia—recognized the existence of ethnic groups. Units of local government . . . were created . . . designed to coincide as closely as possible with the territory occupied by the most numerous ethnicities.

(244)

Rebirth of Nationalism in Eastern Europe

The breakup of the Soviet Union and Yugoslavia has given more numerous ethnicities the opportunity to organize nation-states. But the less numerous ethnicities still find themselves existing as minorities in multinational states, or divided among more than one of the new states. Especially severe problems have occurred in the Balkans. Bulgaria's Turkish minority pressed for more rights, including permission to teach the Turkish language as an optional subject in school. But many Bulgarians opposed these efforts. The Soviet Union, Yugoslavia, and Czechoslovakia were dismantled . . . largely because minority ethnicities opposed the long-standing dominance of the most numerous ones in each country. Local government units . . . made peaceful transitions into independent countries—as long as their boundaries corresponded reasonably well with the territory occupied by a clearly defined ethnicity. The relatively close coincidence between the boundaries of the Slovene ethnic group and the country of Slovenia has promoted the country's relative peace and stability, compared to other former Yugoslavian republics. Sovereignty has brought difficulties in converting from Communist economic systems and fitting into the global economy (see Chapters 9 and 11). But . . . problems of economic reform are minor compared to the conflicts . . . where nation-states could not be created.

Key Issue 3. Why Do Ethnicities Clash?
- **Ethnic competition to dominate nationality**
- **Dividing ethnicities among more than one state**

Ethnic Competition to Dominate Nationality
Sub-Saharan Africa has been a region especially plagued by conflicts among ethnic groups competing to become dominant within the various countries. The Horn of Africa and central Africa are the two areas . . . where conflicts . . . have been particularly complex and brutal.

Ethnic Competition in the Horn of Africa
The Horn of Africa encompasses the countries of Djibouti, Ethiopia, Eritrea, and Somalia. Especially severe problems have been found in Ethiopia, Eritrea, and Somalia, as well as the neighboring country of Sudan.

Ethiopia and Eritrea. Eritrea, located along the Red Sea, became an Italian colony in 1890. Ethiopia, an independent country for more than 2,000 years, was captured by Italy during the 1930s. After World War II, Ethiopia regained its independence, and the United Nations awarded Eritrea to Ethiopia. Ethiopia dissolved the Eritrean legislature and banned the use of Tigrinya, Eritrea's major local language. The Eritreans rebelled, beginning a 30-year fight for independence (1961–1991). In 1991 Eritrean rebels defeated the Ethiopian army, and in 1993 Eritrea became an independent state. But war between Ethiopia and Eritrea flared up again in 1998 because of disputes over the location of the border. Ethiopia defeated Eritrea in 2000 and took possession of the disputed areas.

(245)
Even with the loss of Eritrea, Ethiopia remained a complex multi-ethnic state. From the late nineteenth century until the 1990s, Ethiopia was controlled by the Amharas, who are Christians. After the government defeat in the early 1990s, power passed to a combination of ethnic groups. Eritrea has nine major ethnic groups.

Sudan. In Sudan a civil war has raged since the 1980s between two ethnicities, the black Christian and animist rebels in the southern provinces and the Arab Muslim-dominated government forces in the north. The black southerners have been resisting government attempts to convert the country from a multi-ethnic society to one nationality tied to Muslim traditions.

Somalia. On the surface, Somalia should face fewer ethnic divisions than its neighbors in the Horn of Africa. Somalis are overwhelmingly Sunni Muslims and speak Somali. Somalia contains six major ethnic groups known as clans. Traditionally, the six major clans occupied different portions of Somalia.

(246)
With the collapse of a national government in Somalia, various clans and sub-clans claimed control over portions of the country. In 1992, after an estimated 300,000 people . . . died from famine and from warfare between clans, the United States sent several thousand troops to Somalia . . . to protect

delivery of food . . . and to reduce the number of weapons in the hands of the clan and sub-clan armies. After peace talks among the clans collapsed in 1994, U.S. troops withdrew.

Ethnic Competition in Lebanon
Lebanon has been severely damaged by fighting among religious factions since the 1970s. The precise distribution of religions in Lebanon is unknown, because no census has been taken since 1932. Current estimate is about 60 percent Muslim, 30 percent Christian, and 10 percent other. About 7 percent of the population is Druze. The Druze religion combines elements of Islam and Christianity.

When Lebanon became independent in 1943, the constitution required that each religion be represented in the Chamber of Deputies according to its percentage in the 1932 census. By unwritten convention, the president of Lebanon was a Maronite Christian, the premier a Sunni Muslim, the speaker of the Chamber of Deputies a Shiite Muslim, and the foreign minister a Greek Orthodox Christian. Other cabinet members and civil servants were similarly apportioned among the various faiths. Lebanon's religious groups have tended to live in different regions of the country. Maronites are concentrated in the west central part, Sunnis in the northwest, and Shiites in the south and east.

When the governmental system was created, Christians constituted a majority and controlled the country's main businesses, but as the Muslims became the majority, they demanded political and economic equality. A civil war broke out in 1975, and each religious group formed a private army or militia to guard its territory. Syria, Israel, and the United States sent troops into Lebanon at various points to try to restore peace. Most of Lebanon is now controlled by Syria, which has a historical claim over the territory.

(247)
Dividing Ethnicities among More Than One State
Newly independent countries were often created to separate two ethnicities. However, two ethnicities can rarely be segregated completely.

Dividing Ethnicities in South Asia
When the British ended their colonial rule of the Indian subcontinent in 1947, they divided the colony into two irregularly shaped countries: India and Pakistan. The basis for separating West and East Pakistan from India was ethnicity. Antagonism between the two religious groups was so great that the British decided to place the Hindus and Muslims in separate states.

(248)
Forced Migration. The partition of South Asia into two states resulted in massive migration, because the two boundaries did not correspond precisely to the territory inhabited by the two ethnicities. Hindus in Pakistan and Muslims in India were killed attempting to reach the other side of the new border by people from the rival religion.

Ethnic Disputes. Pakistan and India never agreed on the location of the boundary separating the two countries in the northern region of Kashmir. Since 1972 the two countries have maintained a "line of control" through the region. Muslims, who comprise a majority in both portions, have fought a guerrilla war to secure reunification of Kashmir, either as part of Pakistan or as an independent country. India's religious unrest is further complicated by the presence of 19 million Sikhs, who have long resented that they were not given their own independent country when India was partitioned (see Chapter 6). Sikhs comprise a majority in the Indian state of Punjab. Sikh extremists have fought for more control over the Punjab or even complete independence from India.

(249)
Dividing Sri Lanka Among Ethnicities
Sri Lanka (formerly Ceylon), an island country of 20 million inhabitants off the Indian coast, has been torn by fighting between the Sinhalese and the Tamils. Sinhalese, who comprise 74 percent of Sri Lanka's population, migrated from northern India in the fifth century B.C., occupying the southern

portion of the island. Tamils—18 percent of Sri Lanka's population— migrated across the narrow 80-kilometer-wide (50-mile) Palk Strait from India beginning in the third century B.C. and occupied the northern part of the island. The dispute between Sri Lanka's two ethnicities extends back more than 2,000 years but was suppressed during 300 years of European control. Since independence in 1948, Sinhalese have dominated. Tamils have received support, from Tamils living in other countries, for a rebellion that began in 1983.

(250)
Key Issue 4. What Is Ethnic Cleansing?
   - **Ethnic cleansing in Yugoslavia**
   - **Ethnic cleansing in Central Africa**

Throughout history, ethnic groups have been forced to flee from other ethnic groups' more powerful armies. The largest level of forced migration came during (and after) World War II. The scale of forced migration during (and after) World War II has not been repeated, but in the 1990s a new term—"ethnic cleansing"—was invented to describe new practices by ethnic groups against other ethnic groups. **Ethnic cleansing** is a process in which a more powerful ethnic group forcibly removes a less powerful one in order to create an ethnically homogeneous region.

Ethnic Cleansing in Yugoslavia
Ethnic cleansing in former Yugoslavia is part of a complex pattern of ethnic diversity in the region of southeastern Europe known as the Balkan Peninsula. The Balkans includes Albania, Bulgaria, Greece, and Romania, as well as several countries that once comprised Yugoslavia.

(251)
Creation of Multi-ethnic Yugoslavia
The Balkan Peninsula has long been a hotbed of unrest, a complex assemblage of ethnicities. Northern portions were incorporated into the Austria-Hungary Empire, whereas southern portions were ruled by the Ottomans. In June 1914 the heir to the throne of Austria-Hungary was assassinated in Sarajevo by a Serb who sought independence for Bosnia. The incident sparked World War I. After World War I the allies created a new country, Yugoslavia, to unite several Balkan ethnicities that spoke similar South Slavic languages. The prefix "Yugo" in the country's name derives from the Slavic word for "south."

Ethnic Diversity in the Former Yugoslavia. Under the long leadership of Josip Broz Tito, who governed Yugoslavia from 1953 until his death in 1980, Yugoslavs liked to repeat a refrain that roughly translates as follows: "Yugoslavia has seven neighbors, six republics, five nationalities, four languages, three religions, two alphabets, and one dinar." The refrain concluded that Yugoslavia had one dinar, the national unit of currency. Despite cultural diversity, according to the refrain, common economic interests kept Yugoslavia's nationalities unified.

Destruction of Multi-Ethnic Yugoslavia
Rivalries among ethnicities resurfaced in Yugoslavia during the 1980s after Tito's death, leading to the breakup of the country in the early 1990s. When Yugoslavia's republics were transformed from local government units into five separate countries, ethnicities fought to redefine the boundaries.

(252)
Ethnic Cleansing in Bosnia. The creation of a viable country proved especially difficult in the case of Bosnia and Herzegovina. Rather than live in an independent multi-ethnic country with a Muslim plurality, Bosnia and Herzegovina's Serbs and Croats fought to unite the portions of the republic that they inhabited with Serbia and Croatia, respectively. Ethnic cleansing by Bosnian Serbs against Bosnian Muslims was especially severe, because much of the territory inhabited by Bosnian Serbs was separated from Serbia by areas with Bosnian Muslim majorities. Accords reached in Dayton, Ohio, in 1996 . . . divided Bosnia and Herzegovina into three regions, one each dominated by the Bosnian

Croats, Muslims, and Serbs. Bosnian Muslims, 44 percent of the population before the ethnic cleansing, got 27 percent of the land.

(253)
Ethnic Cleansing in Kosovo. Despite the loss of Bosnia, Croatia, and Slovenia in the early 1990s, Yugoslavia remained a multi-ethnic country, although dominated by Serbs. Particularly troubling was the province of Kosovo, where ethnic Albanians comprised 90 percent of the population. Serbia had an historical claim to Kosovo, having controlled it between the twelfth and fourteenth centuries. Serbia was given control of Kosovo when Yugoslavia was created in the early twentieth century. Under Tito, ethnic Albanians in Kosovo received administrative autonomy and national identity. As most Serbs emigrated from Kosovo north into Serbia, the percentage of Albanians in Kosovo increased from one half in 1946 to three fourths at the time of Yugoslavia's last formal census in 1981.

(254)
With the breakup of Yugoslavia, Serbia took direct control of Kosovo and launched a campaign of ethnic cleansing of the Albanian majority. Outraged by the ethnic cleansing, the United States and Western European democracies, operating through the North Atlantic Treaty Organization (NATO), launched an air attack against Serbia.

(255)
Balkanization. A century ago, the term **Balkanized** was widely used to describe a small geographic area that could not successfully be organized into one or more stable states because it was inhabited by many ethnicities with complex, long-standing antagonisms toward each other. Balkanization directly led to World War I. At the end of the twentieth century—after two world wars and the rise and fall of communism—the Balkans have once again become Balkanized. If peace comes to the Balkans, it will be because in a tragic way ethnic cleansing "worked." Millions of people were rounded up and killed or forced to migrate . . . . Ethnic homogeneity may be the price of peace in areas that once were multi-ethnic.

Central Africa
Long-standing conflicts between two ethnic groups, the Hutus and Tutsis, lie at the heart of a series of wars in central Africa. The Hutus were settled farmers. The Tutsi were cattle herders who migrated . . . . from the Rift Valley of western Kenya beginning 400 years ago. The Tutsi took control of the kingdom of Rwanda and turned the Hutu into their serfs. Under German and Belgian control, differences between the two ethnicities were reinforced. Shortly before Rwanda gained its independence in 1962, Hutus killed or ethnically cleansed most of the Tutsis out of fear that the Tutsis would seize control of the newly independent country. In 1994 children of the ethnically cleansed Tutsis, most of whom lived in neighboring Uganda, poured back into Rwanda, defeated the Hutu army, and killed a half-million Hutus, while suffering a half-million casualties of their own. Three million of the country's 7 million Hutus fled to Zaire, Tanzania, Uganda, and Burundi. The conflict between Hutus and Tutsis spilled into neighboring countries of central Africa, especially the Democratic Republic of Congo. Tutsis were instrumental in the successful overthrow of the Congo's longtime president, Joseph Mobutu, in 1997, replacing him with Laurent Kabila. But Tutsis soon split with Kabila and led a rebellion that gained control of the eastern half of the Congo. Armies from Angola, Namibia, Zimbabwe, and other neighboring countries came to Kabila's aid.

Key Terms

Apartheid (p.235)
Balkanization (p.255)
Balkanized (p.255)
Blockbusting (p.235)
Centripetal force (p.239)
Ethnic cleansing (p.250)
Ethnicity (p.227)
Multi-ethnic state (p.239)

Multinational state (p.239)

Nationalism (p.238)
Nationality (p.237)
Nation-state (p.237)
Race (p.227)
Racism (p.233)
Racist (p.233)
Self-determination (p.237)

Sharecropper (p.231)   Triangular slave trade (p.230)

Chapter 8. Political Geography

Today human geographers emphasize a thematic approach, concerned with the location of activities in the world, the reasons for particular spatial distributions, and the significance of the arrangements. Political geographers study how people have organized Earth's land surface into countries and alliances, reasons underlying the observed arrangements, and the conflicts that result from the organization.

Key Issues
1. Where are states located?
2. Why do boundaries between states cause problems?
3. Why do states cooperate with each other?
4. Why has terrorism increased?

(263)

With the end of the Cold War in the 1990s, the global political landscape changed fundamentally. Geographic concepts help us to understand this changing political organization of Earth's surface. We can also use geographic methods to examine the causes of political change. Boundary lines are not painted on Earth, but they might as well be, for these national divisions are very real. To many, national boundaries are more meaningful than natural features. In the post–Cold War era, the familiar division of the world into countries or states is crumbling. Between the mid-1940s and the late 1980s two superpowers—the United States and the Soviet Union—essentially "ruled" the world. But the United States is less dominant in the political landscape of the twenty-first century, and the Soviet Union no longer exists. Today globalization means more connections among states. Power is exercised through connections among states created primarily for economic cooperation. Despite (or perhaps because of) greater global political cooperation, local diversity has increased in political affairs, as individual cultural groups demand more control over the territory they inhabit.

Key Issue 1. Where Are States Located?
- **Problems of defining states**
- **Development of the state concept**

As recently as a half century ago, the world contained only about 50 countries, compared to nearly 200 today. A **state** is an area organized into a political unit and ruled by an established government that has control over its internal and foreign affairs. The term country is a synonym for state. The 50 states of the United States are subdivisions within a single state: the United States of America. Antarctica is the only large landmass on Earth's surface that is not part of a state. Several states . . . claim portions of Antarctica. The United States, Russia, and a number of other states do not recognize the claims of any country to Antarctica. The Treaty of Antarctica, signed in 1959 and renewed in 1991, provides a legal framework for managing Antarctica.

Problems of Defining States
There is some disagreement about the actual number of sovereign states. Among places that test the definition of a state are Korea, China, and Western Sahara (Sahrawi Republic).

Korea: One State or Two?
A colony of Japan for many years, Korea was divided into two occupation zones by the United States and former Soviet Union after they defeated Japan in World War II. Both Korean governments are committed to reuniting the country into one sovereign state. Meanwhile, in 1992, North Korea and South Korea were admitted to the United Nations as separate countries.

(264)
China and Taiwan: One State or Two?
According to China's government officials, Taiwan is not a separate sovereign state but is a part of China. Until 1999 the government of Taiwan agreed. This confusing situation arose from a civil war.

After losing, nationalist leaders in 1949 fled to the island of Taiwan, 200 kilometers (120 miles) off the Chinese coast . . . (and) proclaimed that they were still the legitimate rulers of the entire country of China.

(265)
Most other governments in the world consider China and Taiwan as two separate and sovereign states. Taiwan's president announced in 1999 that Taiwan would also regard itself as a sovereign independent state.

The United States had supported the Nationalists during the civil war . . . (and) continued to regard the Nationalists as the official government of China until 1971, when U.S. policy finally changed and the United Nations voted to transfer China's seat from the Nationalists to the Communists.

Western Sahara (Sahrawi Republic). The Sahrawi Arab Democratic Republic is considered by most African countries as a sovereign state. Morocco, however, controls the territory, which it calls Western Sahara. The United Nations is sponsoring a referendum for the residents of Western Sahara to decide whether they want independence or want to continue to be part of Morocco.

(266)
Varying Size of States
The land area occupied by the states of the world varies considerably. The largest state is Russia, which encompasses 17.1 million square kilometers (6.6 million square miles), or 11 percent of the world's entire land area. (Five) other states with more than 5 million square kilometers (2 million square miles) include China, Canada, United States, Brazil, and Australia.

At the other extreme are about two dozen **microstates**, which are states with very small land areas. The smallest microstate in the United Nations—Monaco—encompasses only 1.5 square kilometers (0.6 square miles). Many of these are islands, which explains both their small size and sovereignty.

Development of the State Concept
The concept of dividing the world into a collection of independent states is recent. Prior to the 1800s, Earth's surface was organized in other ways, such as city-states, empires, and tribes. Much of Earth's surface consisted of unorganized territory.

(267)
Ancient and Medieval States. The modern movement to divide the world into states originated in Europe. However, the development of states can be traced to . . . the Fertile Crescent.

Ancient States. Situated at the crossroads of Europe, Asia, and Africa, the Fertile Crescent was a center for land and sea communications in ancient times.

The first states to evolve in Mesopotamia were known as city-states. A **city-state** is a sovereign state that comprises a town and the surrounding countryside. Periodically, one city or tribe in Mesopotamia would gain military dominance over the others and form an empire. Meanwhile, the state of Egypt emerged as a separate empire at the western end of the Fertile Crescent . . . (in a) long, narrow region along the banks of the Nile River. Egypt's empire lasted from approximately 3000 B.C. until the fourth century B.C.

Early European States. Political unity in the ancient world reached its height with the establishment of the Roman Empire, which controlled most of Europe, North Africa, and Southwest Asia, from modern-day Spain to Iran and from Egypt to England.

The Roman Empire collapsed in the fifth century A.D. after a series of attacks by people living on its frontiers, as well as internal disputes. The European portion of the Roman Empire was fragmented into a large number of estates. Victorious nobles seized control of defeated rivals' estates. Most people were forced to live on an estate, working and fighting for the benefit of the noble. Beginning about the year 1100, a handful of powerful kings emerged as rulers over large numbers of estates. The consolidation . . . formed the basis for the development of such modern Western European states as England, France, and Spain. Central Europe . . . remained fragmented . . . until the nineteenth century.

Colonies
A colony is a territory that is legally tied to a sovereign state rather than being completely independent.

(268)
Colonialism. European states came to control much of the world through colonialism. European states established colonies . . . for three basic reasons: to promote Christianity; (to) provide resources; (and) to indicate relative power. The three motives can be summarized as God, gold, and glory. The (global) colonial era began in the 1400s. The European states eventually lost most of their Western Hemisphere colonies . . . then turned their attention to Africa and Asia.

The United Kingdom assembled by far the largest colonial empire, (with) colonies on every continent. France had the second-largest overseas territory, although its colonies were concentrated in West Africa and Southeast Asia. Both the British and the French also took control of a large number of strategic islands.

Portugal, Spain, Germany, Italy, Denmark, the Netherlands, and Belgium all established colonies outside Europe, but they controlled less territory than the British and French. Germany tried to compete with Britain and France by obtaining African colonies that would interfere with communications in the rival European holdings.

Colonial Practices. The colonial practices of European states varied. France attempted to assimilate its colonies into French culture. The British created different government structures and policies for various territories of their empire. This decentralized approach helped to protect the diverse cultures. Most African and Asian colonies became independent after World War II.

(269)
The Few Remaining Colonies. At one time, colonies were widespread over Earth's surface, but today only a handful remains. Nearly all are islands in the Pacific Ocean or Caribbean Sea. The most populous remaining colony is Puerto Rico, which is a Commonwealth of the United States. Its 4 million residents are citizens of the United States. The world's least populated colony is Pitcairn Island . . . settled in 1790 by British mutineers.

(270)
Key Issue 2. Where Are Boundaries Drawn between States?
- **Shapes of states**
- **Types of boundaries**
- **Internal organization of states**

A state is separated from its neighbors by a boundary. Boundaries result from a combination of natural physical features (such as rivers, deserts, mountains) and cultural features (such as language and religion). Boundaries interest geographers because the process of selecting their location is frequently difficult. Boundaries . . . also commonly generate conflict, both within a country and with its neighbors.

(271)
Shapes of States
The shape of a state . . . affects the potential for communications and conflict with neighbors, . . . can influence the ease or difficulty of internal administration, and can affect social unity.

Five Basic Shapes
Countries have one of five basic shapes: compact, prorupted, elongated, fragmented, and perforated.

Compact States: Efficient. In a **compact state**, the distance from the center to any boundary does not vary significantly. Compactness is a beneficial characteristic for most smaller states, because good communications can be more easily established to all regions.

Prorupted States: Access or Disruption. An otherwise compact state with a large projecting extension is a **prorupted state**. Proruptions are created for two principal reasons. First, a proruption can provide a state with access to a resource, such as water. Proruptions can also separate two states that otherwise would share a boundary.

(272)
Elongated States: Potential Isolation. There are a handful of elongated states, or states with a long and narrow shape. The best example is Chile. A less extreme example of an elongated state is Italy. Elongated states may suffer from poor internal communications.

(273)
Fragmented States: Problematic. A **fragmented state** includes several discontinuous pieces of territory. There are two kinds of fragmented states: those with areas separated by water, and those separated by an intervening state.

A difficult type of fragmentation occurs if the two pieces of territory are separated by another state. Picture the difficulty of communicating between Alaska and the lower 48 states if Canada were not a friendly neighbor. Perhaps the most intractable fragmentation results from a tiny strip of land in India called Tin Bigha, . . . only 178 meters (about 600 feet) by 85 meters (about 300 feet). For most of the twentieth century, Panama was an example of a fragmented state divided in two parts by the Canal, built in 1914 by the United States.

(274)
Perforated States: South Africa. A state that completely surrounds another one is a **perforated state**. The one good example of a perforated state is South Africa, which completely surrounds the state of Lesotho.

Landlocked States
Lesotho is unique in being completely surrounded by only one state, but it shares an important feature with several other states in southern Africa, as well as in other regions: It is landlocked. The prevalence of **landlocked states** in Africa is a remnant of the colonial era, when Britain and France controlled extensive regions.

Direct access to an ocean is critical to states because it facilitates international trade. To send and receive goods by sea, a landlocked state must arrange to use another country's seaport.

Landlocked States in Southern Africa. Cooperation between landlocked states in southern Africa has been complicated by racial patterns. In the past, the states of southern Africa had to balance their economic dependency on South Africa with their dislike of the country's racial policies.

(275)
Types of Boundaries
Historically, frontiers rather than boundaries separated states. A **frontier** is a zone where no state exercises complete political control. A frontier is a tangible geographic area, whereas a boundary is an infinitely thin, invisible, imaginary line.

A frontier area is either uninhabited or sparsely settled by a few isolated pioneers seeking to live outside organized society. Almost universally, frontiers between states have been replaced by boundaries. The only regions of the world that still have frontiers rather than boundaries are Antarctica and the Arabian Peninsula.

Boundaries are of two types: physical and cultural. Neither type of boundary is better or more "natural," and many boundaries are a combination of both types.

Physical Boundaries
Important physical features on Earth's surface can make good boundaries because they are easily seen, both on a map and on the ground. Three types of physical elements serve as boundaries between states: mountains, deserts, and water.

Mountain Boundaries. Mountains can be effective boundaries if they are difficult to cross . . . (and) because they are rather permanent and usually are sparsely inhabited. Mountains do not always provide for the amicable separation of neighbors. Argentina and Chile agreed to be divided by the crest of the Andes Mountains but could not decide on the precise location of the crest.

Desert Boundaries. Like mountains, deserts are hard to cross and sparsely inhabited. Desert boundaries are common in Africa and Asia.

Water Boundaries. Rivers, lakes, and oceans are the physical features most commonly used as boundaries. Water boundaries are especially common in East Africa.

(276)
Boundaries are typically in the middle of the water, although the boundary between Malawi and Tanzania follows the north shore of Lake Malawi (Lake Nyasa). Again, the boundaries result from nineteenth-century colonial practices: Malawi was a British colony, whereas Tanzania was German. Water boundaries can offer good protection against attack from another state, because an invading state must . . . secure a landing spot. The state being invaded can concentrate its defense at the landing point. The use of water as boundaries between states can cause difficulties, though. One problem is that the precise position of the water may change over time. Rivers, in particular, can slowly change their course.

Ocean boundaries also cause problems because states generally claim that the boundary lies not at the coastline but out at sea. The reasons are for defense and for control of valuable fishing industries.

Cultural Boundaries
The boundaries between some states coincide with differences in ethnicity. Other cultural boundaries are drawn according to geometry; they simply are straight lines drawn on a map.

Geometric Boundaries. Part of the northern U.S. boundary with Canada is a 2,100-kilometer (1,300-mile) straight line (more precisely, an arc) along 49° north latitude, . . . established in 1846 by a treaty between the United States and Great Britain, which still controlled Canada. The United States and Canada share an additional 1,100-kilometer (700-mile) geometric boundary between Alaska and the Yukon Territory along the north-south arc of 14° west longitude.

(277)
The 1,000-kilometer (600-mile) boundary between Chad and Libya is a straight line drawn across the desert in 1899 by the French and British. Subsequent actions by European countries created confusion over the boundary.

Religious Boundaries. Religious differences often coincide with boundaries between states, but in only a few cases has religion been used to select the actual boundary line. The most notable example was in South Asia, when the British partitioned India into two states on the basis of religion. Religion was also used to some extent to draw the boundary between two states on the island of Eire (Ireland).

Language Boundaries. Language is an important cultural characteristic for drawing boundaries, especially in Europe. By global standards, European languages have substantial literary traditions and formal rules of grammar and spelling.

The French language was a major element in the development of France as a unified state in the seventeenth century. In the nineteenth century, Italy and Germany also emerged as states that unified the speakers of particular languages. The movement to identify nationalities on the basis of language spread throughout Europe in the twentieth century. After World War I,
. . . the Versailles Peace Conference (met) to redraw the map of Europe. The geographer Isaiah Bowman played a major role in the decisions. Language was the most important criterion . . . used to create new states . . . and to adjust the boundaries of existing ones. The conference was particularly concerned with Eastern and Southern Europe, regions long troubled by political instability and conflict. Although the boundaries imposed by the Versailles conference on the basis of language were adjusted somewhat after World War II, they proved to be relatively stable, and peace ensued for several decades. However, during the 1990s, the map of Europe drawn at Versailles in 1919 collapsed.

Cyprus' "Green Line" Boundary. Cyprus, the third-largest island in the Mediterranean Sea, contains two nationalities: Greek and Turkish. When Cyprus gained independence from Britain in 1960, its constitution guaranteed the Turkish minority a substantial share of elected offices and control over its own education, religion, and culture.

(278)
Cyprus has never peacefully integrated the Greek and Turkish nationalities. In 1974 several Greek Cypriot military officers who favored unification of Cyprus with Greece seized control of the government. Turkey invaded Cyprus to protect the Turkish Cypriot minority. Traditionally, the Greek and Turkish Cypriots mingled, but after the coup and invasion, the two nationalities became geographically isolated. The Turkish sector declared itself the independent Turkish Republic of Northern Cyprus in 1983, but only Turkey recognizes it as a separate state. A buffer zone patrolled by U.N. soldiers stretches across the entire island to prevent Greeks and Turks from crossing. Some cooperation continues between sectors: The Turks supply the Greek side with water and in return receive electricity.

The United Nations put together a plan to reunite the two portions of Cyprus into a single country with considerable autonomy for each side. Adding pressure to the reunification movement, the European Union agreed to accept the entire island of Cyprus as a member in 2004. The Turkish Cypriots opened the borders between the two sides in 2003. However, many generations of distrust made resolution and implementation of a final agreement difficult.

Boundaries inside States
Within countries, local government boundaries are sometimes drawn to separate different nationalities or ethnicities. They are also drawn sometimes to provide advantage to a political party.

Unitary and Federal States
In the face of increasing demands by ethnicities for more self-determination, states have restructured their governments to transfer some authority from the national government to local government units.

The governments of states are organized according to one of two approaches: the unitary system or the federal system. The **unitary state** places most power in the hands of central government officials, whereas the **federal state** allocates strong power to units of local government within the country.

(279)
In principle, the unitary government system works best in nation-states characterized by few internal cultural differences and a strong sense of national unity. Unitary states are especially common in Europe. In reality, multinational states often have adopted unitary systems, so that the values of one

nationality can be imposed on others. In a federal state, such as the United States, local governments possess more authority to adopt their own laws. Multinational states may adopt a federal system of government to empower different nationalities, especially if they live in separate regions of the country.

The federal system is also more suitable for very large states because the national capital may be too remote to provide effective control over isolated regions. The size of the state is not always an accurate predictor of the form of government.

Trend toward Federal Government
In recent years there has been a strong global trend toward federal government.

France: Curbing a Unitary Government. A good example of a nation-state, France has a long tradition of unitary government in which a very strong national government dominates local government decisions. Their basic local government unit is the *département*. A second tier of local government in France is the *commune*. The French government has granted additional legal powers to the departments and communes in recent years. In addition, 22 regional councils that previously held minimal authority have been converted into full-fledged local government units.

Poland: A New Federal Government. Poland switched from a unitary to a federal system after control of the national government was wrested from the Communists. Under the Communists' unitary system, local governments held no legal authority. Poland's 1989 constitution called for a peaceful revolution: creation of 2,400 new municipalities, to be headed by directly elected officials. To these municipalities, the national government turned over ownership of housing, water supplies, transportation systems, and other publicly owned structures. Businesses owned by the national government . . . were either turned over to the municipalities or converted into private enterprises.

The transition to a federal system of government proved difficult in Poland and other Eastern European countries. The first task for many newly elected councilors was to attend a training course in how to govern.

(280)
Electoral Geography
The boundaries separating legislative districts within the United States and other countries are redrawn periodically to ensure that each district has approximately the same population. Boundaries must be redrawn because migration inevitably results in some districts gaining population, whereas others are losing.

The job of redrawing boundaries in most European countries is entrusted to independent commissions. In most U.S. states the job of redrawing boundaries is entrusted to the state legislature. The process of redrawing legislative boundaries for the purpose of benefiting the party in power is called gerrymandering. The term gerrymandering was named for Elbridge Gerry (1744–1814), governor of Massachusetts (1810–12) and vice president of the United States (1813–14).

Types of Gerrymandering. Gerrymandering takes three forms. "Wasted vote" spreads opposition supporters across many districts but in the minority. "Excess vote" concentrates opposition supporters into a few districts. "Stacked vote" links distant areas of like-minded voters through oddly shaped boundaries.

"Stacked Vote" Gerrymandering. Recent gerrymandering in the United States has been primarily "stacked vote."

(281)

"Stacked vote" gerrymandering has been especially attractive to create districts inclined to elect ethnic minorities. Through gerrymandering, only about one-tenth of Congressional seats are competitive, making a shift of more than a few seats increasingly improbable from one election to another in the United States.

Key Issue 3. Why Do States Cooperate with Each Other?
- **Political and military cooperation**
- **Economic cooperation**

Chapter 7 illustrated examples of threats to the survival of states from the trend toward local diversity. In a number of cases, the inability to accommodate the diverse aspirations of ethnicities has led to the breakup of states into smaller ones. The future of the world's current collection of sovereign states is also threatened by the trend toward globalization. States are willingly transferring authority to regional organizations, established primarily for economic cooperation.

Political and Military Cooperation
During the Cold War era (late 1940s until early 1990s) most states joined the United Nations, as well as regional organizations, . . . established primarily to prevent a third world war.

The United Nations
When established in 1945, the United Nations comprised 49 states, but membership grew to 189 in 2000, making it a truly global institution. Switzerland and Taiwan are the only two populous countries that are not in the United Nations. Taiwan resigned when the United Nations voted to admit the People's Republic of China in 1971. The number of countries in the United Nations has increased rapidly on three occasions: 1955, 1960, and the early 1990s.

U.N. members can vote to establish a peacekeeping force and request states to contribute military forces. During the Cold War era, U.N. peacekeeping efforts were often stymied because any one of the five permanent members of the Security Council . . . could veto the operation. Because it must rely on individual countries to supply troops, the United Nations often lacks enough troops to keep peace effectively. Despite its shortcomings . . . the United Nations represents a forum where, for the first time in history, virtually all states of the world can meet and vote on issues without resorting to war.

(282)
Regional Military Alliances. In addition to joining the United Nations, many states joined regional military alliances after World War II.

Era of Two Superpowers. During the Cold War era, the United States and the Soviet Union were the world's two superpowers. Before then, the world typically contained more than two superpowers. During the Napoleonic Wars in the early 1800s, Europe boasted eight major powers. Before the outbreak of World War I in the early twentieth century, eight great powers again existed. When a large number of states ranked as great powers were of approximately equal strength, no single state could dominate. Instead, major powers joined together to form temporary alliances.

A condition of roughly equal strength between opposing alliances is known as a **balance of power**. Historically, the addition of one or two states to an alliance could tip the balance of power. The British in particular entered alliances to restore the balance of power and prevent any other state from becoming too strong. In contrast, the post–World War II balance of power was bipolar between the United States and the Soviet Union.

Other states lost the ability to tip the scales significantly in favor of one or the other superpower. They were relegated to a new role, that of ally or satellite. Both superpowers repeatedly demonstrated that they would use military force if necessary to prevent an ally from becoming too independent.

Military Cooperation in Europe. After World War II, most European states joined one of two military alliances dominated by the superpowers: NATO or the Warsaw Pact. NATO and the Warsaw Pact were designed to maintain a bipolar balance of power in Europe. In a Europe no longer dominated by military confrontation between two blocs, the Warsaw Pact and NATO became obsolete. Rather than disbanding, NATO expanded its membership in 1997 to include several former Warsaw Pact countries.

(283)
The Organization on Security and Cooperation in Europe (OSCE) has 55 members, including the United States, Canada, and Russia, as well as most European countries. Although the OSCE does not directly command armed forces, it can call upon member states to supply troops if necessary.

Other Regional Organizations. The Organization of American States (OAS) includes all 35 states in the Western Hemisphere. Cuba is a member but was suspended . . . in 1962. The OAS promotes social, cultural, political, and economic links among member states. A similar organization encompassing all countries in Africa is the Organization for African Unity (OAU). Founded in 1963, the OAU has promoted the end of colonialism in Africa. The Commonwealth of Nations includes the United Kingdom and 53 other states that were once British colonies. Commonwealth members seek economic and cultural cooperation.

(284)
Economic Cooperation
The era of a bipolar balance of power formally ended when the Soviet Union was disbanded in 1992. The world has returned to the pattern of more than two superpowers. But the contemporary pattern of global power displays two key differences. The most important elements of state power are increasingly economic rather than military, (and) the leading superpower in the 1990s is not a single state.

European Union
With the decline in the military-oriented alliances, European states increasingly have turned to economic cooperation. Western Europe's most important economic organization is the European Union (formerly known as the European Economic Community, the Common Market, and the European Community).

(286)
When it was established in 1958, the European Union included 6 countries. The European Union has taken on more importance in recent years, as member states seek greater economic and political cooperation. It has removed most barriers to free trade. The introduction of the euro as the common currency in 11 European Union countries eliminates most of the remaining differences in prices, interest rates, and other economic policies within the region. The effect of these actions has been to turn Western Europe into the world's wealthiest market.

Former Communist Countries and the European Union. In 1949 . . . the seven Eastern European states . . . formed an organization for economic cooperation, the Council for Mutual Economic Assistance (COMECON). Cuba, Mongolia, and Vietnam were also members. Like the Warsaw Pact, COMECON disbanded in the early 1990s.

(287)
Joining the European Union in 2004 were eight former Communist Eastern European countries that had made the most progress in converting to market economies: the Czech Republic, Estonia, Hungary, Latvia, Lithuania, Poland, the Slovak Republic, and Slovenia. Also joining in 2004 were the island countries of Cyprus and Malta. Current EU members are wary of admitting a large number of relatively poor Southern and Eastern European countries.

(288)

Key Issue 4. Why Has Terrorism Increased?
- **Terrorism by individuals and organizations**
- **State support for terrorism**

Terrorism is the systematic use of violence by a group in order to intimidate a population or coerce a government into granting its demands. Violence is considered necessary by terrorists to bring widespread publicity to goals and grievances that are not being addressed through peaceful means.

The term *terror* (from the Latin "to frighten") was first applied to the period of the French Revolution between March 1793 and July 1794 known as the Reign of Terror. In modern times, terrorism has been applied to actions by groups operating outside government rather than by official government agencies, although some governments provide military and financial support for terrorists.

Terrorism differs from assassinations and other acts of political violence because terrorist attacks are aimed at ordinary people rather than military targets or political leaders. Average individuals are unintended victims rather than principal targets in most conflicts, whereas a terrorist considers all citizens responsible for the actions being opposed, so therefore equally justified as victims.

Distinguishing terrorism from other acts of political violence can be difficult. For example, if a Palestinian suicide bomber kills several dozen Israeli teenagers in a Jerusalem restaurant, is that an act of terrorism or wartime retaliation against Israeli government policies and army actions? Spokespersons on television make competing claims: Israel's sympathizers denounce the act as a terrorist threat to the country's existence, while advocates of the Palestinian cause argue that long-standing injustices provoked the act. Similarly, Russia claims that Chechen rebels are terrorists, and the British have long claimed that Irish Republican Army rebels are terrorists.

Terrorism by Individuals and Organizations
The United States suffered several terrorist attacks during the late twentieth century. With the exception of the Oklahoma City bombing, which killed 168 people in 1995, Americans generally paid little attention to the attacks and had only a vague notion of who had committed them. It took the attack on the World Trade Center and Pentagon on September 11, 2001, for most Americans to feel threatened by terrorism.

American Terrorists. Some of the terrorists during the 1990s were American citizens operating alone or with a handful of others. Theodore J. Kaczynski, known as the Unabomber, was convicted of killing 3 people and injuring 23 others by sending bombs through the mail during a 17-year period. His targets were mainly academics in technological disciplines and executives in businesses whose actions he considered to be adversely affecting the environment.

Timothy J. McVeigh claimed his terrorist act was provoked by rage against the U.S. government for such actions as the Federal Bureau of Investigation's 51-day siege of the Branch Davidian religious compound near Waco, Texas, culminating with an attack on April 19, 1993, that resulted in 80 deaths.

(289)

Al-Qaeda. Responsible or implicated in most of the anti-U.S. terrorism during the 1990s, as well as the September 11, 2001, attack, was the al-Qaeda network, founded by Osama bin Laden. His father, Mohammed bin Laden, a native of Yemen, established a construction company in Saudi Arabia and became a billionaire through close connections to the royal family. Osama bin Laden, one of about 50 children fathered by Mohammed with several wives, used his several hundred million dollar inheritance to fund al-Qaeda.

Bin Laden moved to Afghanistan during the mid-1980s to support the fight against the Soviet army and the country's Soviet-installed government. Calling the anti-Soviet fight a holy war or *jihad*, bin Laden recruited militant Muslims from Arab countries to join the cause. Al-Qaeda (an Arabic word meaning "the base") was created around 1990 to unite *jihad* fighters in Afghanistan, as well as supporters of bin Laden elsewhere in the Middle East. Most al-Qaeda cell members lived in ordinary society, supporting themselves with jobs, burglary, and credit card fraud. If arrested, members of one cell were not in a position to identify members of other cells.

Bin Laden issued a declaration of war against the United States in 1996, because of U.S. support for Saudi Arabia and Israel. Al-Qaeda's holy war against the United States culminated in simultaneous attacks on the World Trade Center and the Pentagon on September 11,2001. The attacks resulted in nearly 3,000 fatalities.

(290)
Heightened security prevented al-Qaeda from launching another attack in the United States. Instead, al-Qaeda turned to targets in other countries that were lightly guarded.

Al-Qaeda's use of religion to justify attacks posed challenges to both Muslims and Americans. For many Muslims, the challenge was to express disagreement with U.S. policies, yet disavow the terrorist's approach to opposing the United States. For many Americans, the challenge was to distinguish between the peaceful but unfamiliar principles and practice of the world's three quarters of a billion Muslims, and the misuse and abuse of Islam by a handful of terrorists.

State-sponsored Terrorism
States sponsored terrorism at three increasing levels of involvement:
- providing sanctuary for terrorists wanted by other countries;
- supplying weapons, money, and intelligence to terrorists;
- planning attacks using terrorists.

In response to the September 11, 2001, terrorist attack against the United States, the U.S. government accused first Afghanistan, then Iraq, and then Iran of providing at least one of the three levels of state support for terrorists. As part of its war against terrorism, the U.S. government in cooperation with other countries attacked Afghanistan in 2001 and Iraq in 2003 to depose those countries' government leaders considered supporters of terrorism.

A generation earlier, the United States also attacked Libya in retaliation for using terrorists to plan attacks during the 1980s.

Libya
After deposing the King of Libya in a 1969 military coup, Colonel Muammar al-Qaddafi provided terrorists with financial aid to kill his opponents living in exile in Europe. U.S. relations with Libya deteriorated in 1981 after U.S. aircraft shot down attacking Libyan warplanes while conducting exercises over waters the United States considered international but Libya considered inside its territory.

Terrorists sponsored by Libya in 1986 bombed a nightclub in Berlin popular with U.S. military personnel then stationed there, killing two U.S. soldiers (three, including one civilian). In response, U.S. bombers attacked the Libyan cities of Tripoli and Benghazi in a failed attempt to kill Colonel Qaddafi. In 1990, investigators announced that the 1988 destruction of Pan Am Flight 103 over Lockerbie, Scotland, was conducted by Libyan agents. Following eight years of U.N. economic sanctions, Colonel Qaddafi turned over the suspects for a trial that was held in the Netherlands under Scottish law. One of the two was convicted and sentenced to life imprisonment, while the other was acquitted.

Afghanistan
The United States attacked Afghanistan in 2001 when its leaders, known as Taliban, sheltered Osama bin Laden and other al-Qaeda terrorists. Taliban (Arabic for "students of Muslim religious schools") had gained power in Afghanistan in 1995, temporarily suppressing a civil war that had lasted for more than two decades and imposing strict Islamic fundamentalist law on the population.

Afghanistan's civil war began when the King was overthrown by a military coup in 1973 and replaced five years later in a bloody coup by a government sympathetic to the Soviet Union. The Soviet Union sent 115,000 troops to Afghanistan beginning in 1979 after fundamentalist Muslims, known as *mujahedeen*, or "holy warriors," started a rebellion against the pro-Soviet government.

Unable to subdue the *mujahedeen*, the Soviet Union withdrew its troops in 1989, and the Soviet-installed government in Afghanistan collapsed in 1992. After several years of infighting among the factions that had defeated the Soviet Union, Taliban gained control over most of the country.

Six years of Taliban rule came to an end in 2001 following the U.S. invasion. Destroying Taliban was necessary for the United States in order to go after al-Qaeda leaders, including Osama bin Laden, who were living in Afghanistan as guests of the Taliban. Removal of Taliban unleashed a new struggle for control of Afghanistan among the country's many ethnic groups.

(291)
Iraq
The United States attacked Iraq in 2003 in order to remove from power the country's longtime President Saddam Hussein. U.S. officials, supported by the United Kingdom and a few other countries, argued that Hussein was developing weapons of mass destruction that could be turned over to terrorists.

The U.S. confrontation with Iraq predated the war on terrorism. After Iraq invaded neighboring Kuwait in 1990 and attempted to annex it, the U.S.-led coalition launched the 1991 Gulf War known as Operation Desert Storm to drive Iraq out of Kuwait. Although Iraq was defeated in the 1991 Gulf War, Saddam Hussein and the Ba'ath Party remained in power for another dozen years until deposed by the United States in the 2003 war.

In contrast with the 1991 Gulf War, most U.N.-member states did not support the U.S.-led attack in 2003. Most other countries did not view as sufficiently strong the evidence that Iraq still possessed weapons of mass destruction or intended to use them. Hussein's brutal treatment of Iraqis over several decades was widely acknowledged by other countries but not accepted as justification for military action against him.

U.S. assertion that Hussein had close links with al-Qaeda was also challenged by most other countries, as well as by U.S. intelligence agencies. One reason was that Hussein's Ba'ath Party, which ruled Iraq between 1968 and 2003, espoused different principles than the al-Qaeda terrorists.

(292)
Iran
Hostility between the United States and Iran dates from 1979, when a revolution forced abdication of Iran's pro-U.S. Shah Mohammad Reza Pahlavi. Iran and Iraq fought a war between 1980 and 1988 over control of the Shatt al-Arab waterway, formed by the confluence of the Tigris and Euphrates rivers flowing into the Persian Gulf.

Because both Iran and Iraq were major oil producers, the war caused a sharp decline in international oil prices. An estimated 1.5 million died in the war, until it ended when the two countries accepted a UN peace plan. As the United States launched its war on terrorism, Iran was a less immediate target than Afghanistan and Iraq. However, the United States accused Iran of harboring al-Qaeda members and of trying to install a Shiite-dominated government in Iraq after the United States removed Saddam Hussein from power in 2003.

Other states considered by the United States to be state sponsors of terrorism in recent years have included the following:

- Yemen, which served as a base for al-Qaeda cells and sheltered terrorists who attacked the USS *Cole*;
- Sudan, which sheltered Islamic militants, including Osama bin Laden;
- Iran, which had the capability to produce enriched uranium;
- Syria, which was implicated in support of Iranian and Libyan terrorists;
- North Korea, which was developing nuclear weapons capability.

Key Terms

Balance of power (p.282)
Boundary (p.270)
City-state (p.267)
Colonialism (p.268)
Colony (p.267)
Compact state (p.271)
Elongated state (p.272)
Federal state (p.278)
Fragmented state (p.273)
Frontier (p.275)

Gerrymandering (p.280)
Imperialism (p.268)
Landlocked state (p.274)
Microstate (p.266)
Perforated state (p.274)
Prorupted state (p.271)
Sovereignty (p.263)
State (p.263)
Unitary state (p.278)

Chapter 9. Development

What would you think if a very expensive and exclusive resort were built in your neighborhood, and you and your family, who were economically disadvantaged, were expected to work there (for good wages, perhaps) to serve the needs of the vacationers? The world is divided between relatively rich and relatively poor countries. Geographers try to understand the reasons for this division and learn what can be done about it.

Key Issues
1. Why does development vary among countries?
2. Where are more and less developed countries distributed?
3. Where does level of development vary by gender?
4. Why do less developed countries face obstacles to development?

(299)
The second half of the book concentrates on economic rather than cultural elements of human geography. This chapter examines the most fundamental global economic pattern—the division of the world into relatively wealthy regions and relatively poor ones. Subsequent chapters look at the three basic ways that humans earn their living—growing food, manufacturing products, and providing services. Earth's nearly 200 countries can be classified according to their level of development, which is the process of improving the material conditions of people through diffusion of knowledge and technology. The development process is continuous. Every *place* lies at some point along a continuum of development. A **more developed country** (abbreviated **MDC**) . . . has progressed further along the development continuum. A country in an earlier stage is a **less developed country (LDC)**, although many analysts prefer the term **developing country**. More developed countries cluster in some *spaces*, and less developed countries cluster in others. A number of economic, social, and demographic indicators distinguish more and less developed regions. For more developed regions, the economic challenge is to maintain a high level of development at the new *scale*. For less developed countries, the challenge is to find *connections* to the global economy that take advantage of local skills and resources.

Key Issue 1. Why Does Development Vary Among Countries?
*   **Economic indicators of development**
*   **Social indicators of development**
*   **Demographic indicators of development**

The Human Development Index (HDI), created by the United Nations, recognizes that a country's level of development is a function of all three of these factors: *economic, social, and demographic.*

Four factors are combined to produce a country's HDI. These factors include one economic factor, two social factors, and one demographic factor: gross domestic product per capita, the literacy rate, amount of education, and life expectancy. The highest HDI possible is 1.0, or 100 percent. The country with the highest HDI most years has been Canada, at 0.932 in 1997. Japan had a higher HDI than Canada for a few years. The United States has never ranked first. The two dozen lowest-ranking countries are typically in sub-Saharan Africa.

Economic Indicators of Development
The United Nation's HDI includes one economic indicator of development: gross domestic product per capita. Four other economic indicators distinguish more developed from less developed countries—economic structure, worker productivity, access to raw materials, and availability of consumer goods.

Gross Domestic Product Per Capita
The typical worker receives $10 to $15 per hour in more developed countries, compared to less than $0.50 per hour in less developed ones.

(300)

Per capita income is a difficult figure to obtain. Geographers substitute per capita gross domestic product, a more readily available indicator, dividing the GDP by total population. The gross domestic product (GDP) is the value of the total output of goods and services produced in a country, normally during a year. The gross national product (GNP) is similar to the GDP, except that it (GNP) includes income that people earn abroad.

Annual per capita GDP exceeds $20,000 in more developed countries, compared to about $1,000 in less developed countries. The lowest per capita GDPs are found in sub-Saharan Africa, South Asia, and Southeast Asia. The gap in per capita GDP between more and less developed countries has been widening during the past two decades. Many African and Latin American countries have actually experienced declines in per capita GDP.

Per capita GDP—or, for that matter, any other single indicator—cannot measure perfectly the level of a country's development. Few people are starving in less developed countries. About one eighth of the U.S. population is officially classified as poor. Per capita GDP measures average (mean) wealth, not its distribution

Types of Jobs

Average per capita income is higher in MDCs because people typically earn their living by different means than in LDCs. Jobs fall into three categories: primary (including agriculture), secondary (including manufacturing), and tertiary (including services). Workers in the **primary sector** directly extract materials from Earth. The **secondary sector** includes manufacturers.

(301)

The **tertiary** sector involves the provision of goods and services, retailing, banking, law, education, and government.

The distribution of workers among primary, secondary, and tertiary sectors varies sharply between more and less developed countries. A high percentage of agricultural workers in a country indicates that most of its people are spending their days producing food for their own survival. Freed from the task of growing their own food, most people in a more developed country can contribute to an increase in the national wealth by working in the secondary and tertiary sectors. Within MDCs the number of jobs has decreased in the primary and secondary sectors and increased in the tertiary.

(302)

Productivity

Productivity is the value of a particular product compared to the amount of labor needed to make it. Workers in more developed countries produce more with less effort because they have access to more machines, tools, and equipment to perform much of the work. Productivity can be measured by the **value added** per worker, the gross value of the product minus the costs of raw materials and energy.

Raw Materials

Development requires access to raw materials, such as minerals and trees, which can be fashioned into useful products. It also requires energy to operate the factories. The United Kingdom, the first country to develop in the eighteenth century, had abundant supplies of coal and iron ore, used to make steel for tools. European countries took advantage of domestic coal and iron ore to promote industrial development during the nineteenth century. As they ran short of many raw materials, European countries began to import them. The international flow of raw materials sustained development in Europe but retarded it in Africa and Asia. Most former colonies still export raw materials and import finished goods and services. The LDCs that possess energy resources, especially petroleum, have been able to use revenues to finance development. Prices for other raw materials, such as cotton and copper, have fallen because of excessive global supply and declining industrial demand. A country with abundant resources has a better chance of developing. Yet some countries that lack resources—such as Japan, Singapore, South Korea, and Switzerland—have developed through world trade.

Consumer Goods

Part of the wealth generated in more developed countries goes for essential goods and services (food, clothing, and shelter). But the rest is available for consumer goods and services. The wealth used to buy "nonessentials" promotes expansion. Among the thousands of things that consumers buy, three are particularly good indicators of a society's development: motor vehicles, telephones, and televisions. In MDCs the ratio of people to motor vehicles, telephones, and televisions is approaching 1:1. In other words, more developed countries contain nearly one motor vehicle, telephone, and television set for each citizen. The motor vehicle, telephone, and television all play important economic roles.

In contrast, in less developed countries, these products do not play a central role in daily life. The number of individuals per telephone and motor vehicle exceeds 100 in most LDCs. The number of persons per television set varies widely. The variation reflects the rapid diffusion of television in recent years in LDCs. Most people in LDCs are familiar with these consumer goods, even though they cannot afford them. The minority who have these goods may include government officials, landowners, and other elites, whereas the majority who are denied access to these goods may provoke political unrest. In many LDCs the "haves" are concentrated in urban areas; the "have-nots" live in the countryside.

(304)
Motor vehicles, telephones, and televisions also contribute to social and cultural elements of development. As a result of greater exposure to cultural diversity, people in developed countries display different social characteristics from people in LDCs.

Social Indicators of Development

More developed countries use part of their greater wealth to provide schools, hospitals, and welfare services. In turn, this well-educated, healthy, and secure population can be more economically productive.

Education and Literacy

In general, the higher the level of development, the greater are both the quantity and the quality of a country's education. Quantity is the average number of school years attended. The quality is measured in two ways—student/ teacher ratio and literacy rate. The average pupil attends school for about 10 years in more developed countries, compared to only a couple of years in LDCs. The student/teacher ratio is twice as high in less developed countries as in more developed ones.
The **literacy rate** is the percentage of a country's people who can read and write. It exceeds 95 percent in developed countries, compared to less than one-third in many LDCs.

Health and Welfare

When people get sick, more developed countries possess the resources to care for them. The health of a population is influenced by diet. In less developed countries, most people receive less than the daily minimum allowance of calories and proteins.

(306)
The MDCs use part of their wealth to protect people who, for various reasons, are unable to work. More developed countries are hard-pressed to maintain their current levels of public assistance. Economic growth has slowed, while the percentage of people needing public assistance has increased.

Demographic Indicators of Development

The U.N. HDI utilizes life expectancy as a measure of development. Other demographic characteristics that distinguish more and less developed countries include infant mortality, natural increase, and crude birth rates.

Life Expectancy

Babies born today can expect to live into their early forties in less developed countries compared to their mid-seventies in more developed countries (see Figure 2–11). The gap in life expectancy is greater for females than for males. With longer life expectancies, MDCs have a higher percentage of elderly people who have retired and receive public support.

Infant Mortality Rate

About 90 percent of infants survive . . . in less developed countries, whereas in MDCs more than 99 percent survive. The infant mortality rate is greater in the LDCs for several reasons: . . . malnutrition or lack of medicine . . . (or) poor medical practices.

Natural Increase Rate

The natural increase rate averages more than 2 percent annually in less developed countries and less than 1 percent in more developed ones. Greater natural increase strains a country's ability to provide . . . . services that can make its people healthier and more productive.

Crude Birth Rate

Less developed countries have higher natural increase rates because they have higher crude birth rates. The annual crude birth rate exceeds 40 per 1,000 in many LDCs, compared to less than 15 per 1,000 in MDCs. More developed and less developed countries both have annual crude death rates of about 10 per 1,000. Two reasons account for the lack of difference. First, diffusion of medical technology . . . has eliminated or sharply reduced the incidence of several diseases in less developed countries. Second, MDCs have higher percentages of older people. The mortality rate for women in childbirth is significantly higher in LDCs.

Key Issue 2. Where Are More and Less Developed Countries Distributed?
- **More developed regions**
- **Less developed regions**

The countries of the world can be categorized into nine major regions according to their level of development. These regions also have distinctive demographic and cultural characteristics that have been discussed in earlier chapters. Subsequent chapters will show that the nine major regions also differ . . . in . . . economic characteristics.

(307)
In the Western Hemisphere, two regions (can be distinguished)—Anglo-America (Canada and the United States) and Latin America. Despite the considerable diversity . . . the individual countries within these regions display cultural similarities. Europe can be divided into . . . Western and Eastern, (where) . . . distinctive political experiences have produced different levels of economic development. Asia comprises four major cultural regions: East, South, Southeast, and Southwest. Southwest Asia can be combined with North Africa to form the Middle East region. Africa south of the Sahara comprises the ninth major region.

In addition to those nine major regions, two other important areas can be identified: Japan and the South Pacific. Japan . . . contrast(s) sharply with neighboring states in East Asia. The South Pacific, primarily Australia and New Zealand, . . . is much less populous than the nine major regions. The distribution of more and less developed countries reflects a clear global pattern. Nearly all of the less developed countries lie south of 30° north latitude. The *north-south split* between more and less developed countries shows up clearly in world maps of measures of development.

More Developed Regions

Anglo-America, Western Europe, and Eastern Europe—plus Japan and the South Pacific—are considered more developed.

Anglo-America

Language and religious patterns are less diverse in Anglo-America than in other world regions. Cultural diversity generates some tensions in the region. However, Anglo-America's relative homogeneity reduces the possibility that a large minority will be excluded from participating in the region's economy on the basis of cultural characteristics.

(308)

Anglo-America was once the world's major producer . . . but in the past quarter century Japan, Western Europe, and less developed countries have eroded the region's dominance. Americans remain the leading consumers. The region has adapted relatively successfully to the global economy, in part because it is the leading provider of . . . high-tech services . . . and services that promote use of leisure time. Anglo-America is the world's most important food exporter and the only region that could significantly expand the amount of land devoted to agriculture. Few Americans are farmers, but a large percentage (are) engaged in . . . producing or serving food.

Western Europe

On a global scale, Western Europe displays cultural unity. However, the diversity of individual languages and religious practices has been a longtime source of conflict . . . especially when strong national identities were forged. Competition among Western European nationalities caused many wars.

Since the end of World War II . . . Western Europe has become much more unified. Offsetting the increased cultural unity . . . is . . . migration of Muslims and Hindus . . . in search of jobs. Immigrants are responsible for much of the region's population growth, and they have become scapegoats for the region's economic problems. Within Western Europe the level of development is the world's highest in a core area. Because the region's peripheral areas—southern Italy, Portugal, Spain, and Greece—lag somewhat in development, Western Europe as a whole has a slightly lower development level than Anglo-America. To maintain its high level of development, Western Europe must import food, energy, and minerals. Colonies on every continent . . . supplied many resources needed to foster European economic development. Colonization also diffused Western European languages, religions, and social customs worldwide. Now that most colonies have been granted independence, Western Europeans must buy raw materials from other countries. To pay for their imports, Western Europeans provide high-value goods and services.

The elimination of most economic barriers within the European Union makes Western Europe potentially the world's largest and richest market. Most governments have been willing to sacrifice some economic growth in exchange for protection of existing jobs and social services.

Eastern Europe

Eastern Europe has the dubious distinction of being the only region where the HDI has declined significantly since the United Nations created the index in 1990. In 1990 Eastern Europe . . . had an HDI only slightly behind those of Western Europe and Anglo-America.

Eastern Europe's rapidly declining HDI is a legacy of the region's history of Communist rule. Communist parties . . . achieved rapid development, especially during the 1950s and 1960s. Early Communist theorists . . . believed that communism would triumph in more developed countries. Because few (Eastern European) . . . states had modern industries . . . the Communists had to . . . apply their theories to . . . poor, agricultural societies.

(309)

The Communists promoted development during the 1950s and 1960s through economies directed by government officials rather than private entrepreneurs. In the Soviet Union, for example, a national planning commission called Gosplan developed five-year plans to guide economic development.

The five-year plans featured three main development policies. First—(emphasize) heavy industry, mining, electric power, and transportation. Second—disperse production facilities. (Russia) . . . had been (frequently) invaded from the west. They wanted to reduce the vulnerability of their vital industries to attack, (and) . . . to promote equal development throughout the country. Third—locate manufacturing facilities near sources of raw materials rather than near markets. Soviet planners gave lower priority to producing consumer goods.

Eastern European countries in the 1990s dismantled the economic structure inherited from the Communists. Aside from the desire for freedom, the principal reason that Eastern Europeans rejected communism was that central planning proved to be disastrous at running national economies. For many Eastern Europeans, the most fundamental problem was that by concentrating on basic industry, the Communists neglected consumer products. Although restricted from visiting Western countries, many Eastern Europeans could see on television the much higher level of comfort on the other side of the Iron Curtain. Restructuring to market economies has proved painful in Russia and a number of other Eastern European countries. The Czech Republic, Hungary, and Slovenia have converted more rapidly to market economies, taking advantage of their proximity to the relatively developed core region of Western Europe. Some manufactured goods are being exported to wealthier countries in the West.

(310)
The dismantling of the Communist system led to the breakup of Czechoslovakia, the Soviet Union, and Yugoslavia. Czechs . . . believed that rapid conversion to a market economy would bring long-term benefits. Slovaks wanted to slow the pace of change. Similarly, the Soviet Union and Yugoslavia broke up in part because republics such as Russia and Slovenia preferred more rapid economic change than did Belarus and Serbia. The end of communism in these two countries also unleashed long-suppressed friction among ethnicities. The region's HDI may have declined because of production cutbacks, higher death rates, and other stresses associated with the end of communism. Alternatively, higher mortality and lower wealth may be because the Communists inflated statistics when they were in power. Eastern Europe . . . is classified here as more developed. But the low HDI shows the . . . difficulties in comparing levels of development among regions.

(311)
Japan
Anglo-America and Western Europe share many cultural characteristics. Anglo-America was colonized by European immigrants, so the regions share language, religion, and other political, economic, and cultural traditions. Japan, the third major center of development, has a different cultural tradition.

Japan's development is especially remarkable because it has an extremely unfavorable ratio of population to resources. The country has . . . one of the highest physiological densities. Although Japan is one of the world's leading steel producers, it must import virtually all the coal and iron ore needed for steel production. At first, the Japanese economy developed by taking advantage of the country's one asset, an abundant supply of people willing to work hard for low wages. Having gained a foothold in the global economy by selling low-cost products, Japan then began to specialize in high-quality, high-value products. Japan's dominance was achieved in part by concentrating resources in rigorous educational systems and training programs to create a skilled labor force.

South Pacific
The South Pacific has a relatively high HDI but is much less central to the global economy because of its small number of inhabitants and peripheral location. The HDIs of Australia and New Zealand are comparable to those of other MDCs. The area's remaining people are scattered among sparsely inhabited islands that generally are less developed. Australia and New Zealand share many cultural characteristics with the United Kingdom. Australia and New Zealand are net exporters of food and

other resources, especially to the United Kingdom. Increasingly, their economies are tied to Japan and other Asian countries.

Less Developed Regions

Six regions are classified as less developed. The level of development varies widely among the six regions.

Latin America

Most Latin Americans speak one of two Romance languages—Spanish or Portuguese—and adhere to Roman Catholicism. In reality the region is culturally diverse. A large percentage of the population is descendants of inhabitants living in the region prior to the European conquest, while others trace their ancestors to African slaves. Latin Americans are more likely to live in urban areas than people in other developing regions. The region's population is highly concentrated along the Atlantic Coast. Large areas of interior rain forest are being destroyed to sell the timber or to clear the land for settled agriculture.

The level of development is relatively high along the South Atlantic Coast from Curitiba, Brazil, to Buenos Aires, Argentina. Mexico's development has been aided by proximity to the United States. Development is lower in Central America, several Caribbean islands, and the interior of South America. Overall development in Latin America is hindered by inequitable income distribution. Latin American governments encourage redistribution of land to peasants but do not wish to alienate the large property owners, who generate much of the national wealth.

(312)

East Asia

China, the largest country in East Asia, ranks among the world's poorest. Within a few years China is projected to exceed the United States as the world's largest economy, although the U.S. economy would still be much larger on a per capita basis. Traditionally, most Chinese farmers were forced to pay high rents and turn over a percentage of their crops to a property owner. Exploitation of the country's resources by Europe and Japan further retarded China's development.

China's watershed year was 1949, when the Communist party won a civil war and created the People's Republic of China. To ensure the production and distribution of enough food, the Communist government took control of most agricultural land. In recent years such strict control has been loosened. Individuals again are able to own land and control their own production. Agricultural land must be worked intensively to produce enough food for China's large population. The Chinese government controls the daily lives of the citizenry more than in other countries, and the people have difficulty obtaining some goods. Because of government controls, China has a much lower natural increase rate than other LDCs.

Southeast Asia

Southeast Asia's most populous country, Indonesia, includes 13,667 islands. Nearly two-thirds of the population lives on the island of Java. Other than Indonesia, Southeast Asia's most populous countries are Vietnam, Thailand, and the Philippines.

The region has suffered from a half century of nearly continuous warfare. Japan, the Netherlands, France, and the United Kingdom were all forced to withdraw from colonies. The region's tropical climate limits intensive cultivation of most grains. Economic development is also limited in Southeast Asia by several mountain ranges, active volcanoes, and frequent typhoons. This inhospitable environment traditionally kept population growth low. But the injection of Western medicine and technology resulted in one of the most rapid rates of increase. Rice . . . is exported in large quantities from some countries, such as Thailand and Vietnam, but . . . imported to other countries . . . such as Malaysia and the Philippines.

Because of distinctive vegetation and climate, farmers in Southeast Asia concentrate on harvesting products that are used in manufacturing. Southeast Asia also contains a large percentage of the world's tin as well as some petroleum reserves. Development has been rapid in . . . Thailand, Singapore, Malaysia, and the Philippines. The region (is) a major manufacturer of textiles. Thailand (is) the region's center for . . . automobiles and . . . consumer goods. Economic growth in the region slowed during the past decade. Funds for development were sometimes invested unwisely or stolen by corrupt officials. To restore economic confidence among international investors, Southeast Asian countries have been forced to undertake painful reforms that reduce the people's standard of living.

(313)
Middle East
Much of the Middle East is desert that can sustain only sparse concentrations of plant and animal life. Most products must be imported. Because of petroleum exports, the Middle East is the only one of the nine major world regions that enjoys a trade surplus. Government officials in Middle Eastern states, such as Saudi Arabia and the United Arab Emirates, have used the billions of dollars generated from petroleum sales to finance economic development. Many governments in the region have access to more money than they can use to finance development. However, not every country in the region has abundant petroleum reserves. Development possibilities are limited in countries that lack significant petroleum.

The large gap in per capita income between the petroleum-rich countries and those that lack resources causes great tension in the Middle East. People in poorer states held little sympathy for wealthy Kuwait when Iraq invaded it in 1990. The challenge for many Middle Eastern states is to promote development without abandoning the traditional cultural values of Islam. Many Middle Eastern countries . . . prevent diffusion of financial practices that are considered incompatible with Islamic principles. The low level of literacy among women is the main reason the United Nations considers the development among these petroleum-rich states to be lower than the region's wealth would indicate.

To shed more light on the Middle East's lagging development record, the United Nations uses a team of Arab social scientists to construct an Alternative Human Development Index (AHDI). The AHDI points to three causes in the region's relatively low HDI: lack of political freedom, low education and literacy rates, and lack of opportunities for women.

The region also suffers from serious internal cultural disputes, as discussed in Chapters 6 through 8. Most Middle Eastern states have refused to recognize the existence of Israel. Money that could be used to promote development is diverted to military funding and rebuilding war-damaged structures.

The Middle East has also struggled with terrorism. Very few people endorse acts of violence. On the other hand, sympathy is widespread in the Middle East for those advocating alternatives to U.S.-influenced culture and approaches to development.

South Asia
South Asia includes India, Pakistan, Bangladesh, Sri Lanka, and the small Himalayan states of Nepal and Bhutan.

(314)
The region has the world's second-highest population and second-lowest per capita income. India . . . is the world's leading producer of jute, . . . peanuts, sugarcane, and tea. India has (multiple) mineral reserves. However, the overall ratio of population to resources is unfavorable. India is one of the world's leading rice and wheat producers. The region was a principal beneficiary of the Green Revolution. Agricultural productivity in South Asia also depends on climate. Agricultural output declines sharply if the monsoon rains fail to arrive.

Sub-Saharan Africa
Africa has been allocated to two regions. Countries north of the Sahara Desert share economic and cultural characteristics with the Middle East. South of the desert is called sub-Saharan Africa. Sub-

Saharan Africa has a number of assets. Population density is lower than in any other less developed region. The region contains many (mineral) resources important for economic development. Wealth is comparable to levels found in other LDCs. Despite these assets, sub-Saharan Africa has the least favorable prospect for development. And economic conditions in sub-Saharan Africa have deteriorated in recent years.

Some of the region's economic problems are a legacy of the colonial era. Mining companies and other businesses were established to supply European industries with needed raw materials rather than to promote overall economic development.

(315)
Political problems have also plagued sub-Saharan Africa. European colonies were converted to states without regard for the distribution of ethnicities. The fundamental problem in many countries of sub-Saharan Africa is a dramatic imbalance between the number of inhabitants and the capacity of the land to feed the population.

Key Issue 3. Where Does Level of Development Vary by Gender?
- **Gender-related development index**
- **Gender empowerment**

A country's overall level of development masks inequalities in the status of men and women. The United Nations has not found a single country in the world where its women are treated as well as its men.

To measure the extent of each country's gender inequality, the United Nations has created two indexes. The Gender-related Development Index (GDI) compares the level of development of women with that of both sexes. The Gender Empowerment Measure (GEM) compares the ability of women and men to participate in economic and political decision making.

(316)
Gender-related Development Index
The GDI is constructed in a manner similar to the HDI. The GDI combines the same indicators of development used in the HDI—income, literacy, education, and life expectancy.
The GDI penalizes a country for having a large disparity between the well-being of men and women.

A country with complete gender equality would have a GDI of 1.0. No country has achieved that level. The highest ranking country on the basis of 2001 data was Norway, with a GDI of 0.941. Other countries with relatively high GDIs are in Western Europe and North America. As with the HDI, the United States ranks among the leaders in GDI but is not at the very top. The lowest GDIs are in sub-Saharan Africa.

Economic Indicator of Gender Differences
The average income of women is lower than that of men in every country of the world, both MDCs and LDCs. An income gap of more than $15,000 is typical for MDCs. In LDCs the disparity between male and female income is relatively low in dollar terms but high on a percentage basis.

(317)
Social Indicators of Gender Differences
Women are less likely to attend school in LDCs than in MDCs. The gap is especially high at the secondary school (high school) level.

(318)
In sub-Saharan Africa and the Middle East, fewer than one-third of girls attend school. In contrast, school attendance is nearly universal for both boys and girls in MDCs. In Latin America and much of Asia, boys and girls are equally likely to attend school, but attendance is lower than in MDCs.

In MDCs, literacy is nearly universal among both men and women. In Latin America and Asia, literacy is not universal, but rates are similar for men and women. In sub-Saharan Africa and the Middle East, female literacy is low and substantially lower than for males.

(319)
Demographic Indicator of Gender Differences

The demographic development measure in the GDI—life expectancy—displays a different pattern: the gender gap is greater in more developed countries than in less developed ones. In MDCs, a female baby born today is expected to live several years longer than a male baby, whereas in most LDCs, the gap in life expectancy between females and males is only a year or two. The inability of women to outlive men in LDCs derives primarily from the hazards of childbearing.

Although the status of women is lower than that of men in every country of the world, the United Nations has found that the gap between men and women has been reduced in every country during the past quarter century.

Gender Empowerment

The GDI reflects improvements in the standard of living and well-being of women, whereas the GEM measures the ability of women to participate in the process of achieving those improvements. In every country of the world, both MDCs and LDCs, fewer women than men hold positions of economic and political power, according to the United Nations' GEM scoring system.

The GEM is calculated by combining two indicators of economic power (income and professional jobs) and two indicators of political power (managerial jobs and elected jobs). A country with complete equality of power between men and women would have a GEM score of 1.0.

As with GDI, countries with the highest GEMs are MDCs, especially in North America, Northern Europe, and the South Pacific. Lowest scores are in South Asia and sub-Saharan Africa.

(320)
Economic Indicators of Empowerment

Professional and technical jobs are regarded by the United Nations as those offering women the greatest opportunities for advancement to positions of influence in a country's economy.

More than one-half of professional and technical workers in Northern Europe, as well as in North America, are women. In comparison, fewer than one-fourth of professional and technical jobs are held by women in many LDCs. The United Nations' other key indicator of women's power over economic resources is the share of national income held by women. This is the same indicator included in the GDI.

(321)
Political Indicators of Empowerment

One indicator of the political power of women is the percentage of the country's administrative and managerial jobs they hold. The United Nations considers professional jobs already discussed to be a key measure of economic power, whereas managerial jobs represent the ability to influence the process of decision making. As with other indicators, the percentage of managerial jobs held by women is higher in MDCs than in LDCs.

The other key political indicator of empowerment is percentage of women who are elected to public office. Although more women than men vote in most places, no country has a national parliament or congress with a majority of women.

Every country has a lower GEM than GDI. A higher GDI compared to GEM means that women possess a greater share of a country's resources than power over allocation of those resources.

Key Issue 4. Why Do Less Developed Countries Face Obstacles to Development?
- **Development through self-sufficiency**
- **Development through international trade**
- **Financing development**

The indicators presented in the previous key issues reflect sharp differences in the levels of development of more developed and less developed countries. For many of the indicators the gap between the two is widening rather than narrowing. Annual GDP per capita during the past quarter century has increased by $3,000 in less developed countries, compared to $16,000 in more developed ones. Natural increase has dropped by 20 percent in LDCs but by 83 percent in more developed ones. Infant mortality has dropped by one-half in LDCs but by two-thirds in MDCs.

The one-fifth of the world's people living in more developed countries consumes five-sixths of the world's goods, while Africans consume about 1 percent. Less developed countries chose one of two models to promote development. One approach emphasizes international trade; the other advocates self-sufficiency. Each has important advantages and serious problems.

(322)

Development through Self-Sufficiency. For most of the twentieth century, self-sufficiency, or balanced growth, was the more popular of the development alternatives. The world's two most populous countries, China and India, adopted this strategy, as did most African and Eastern European countries.

Elements of Self-Sufficiency Approach. According to the balanced growth approach, a country should spread investment as equally as possible across all sectors of its economy, and in all regions. Reducing poverty takes precedence over encouraging a few people to become wealthy consumers. The approach nurses fledgling businesses . . . by isolating them from competition of large international corporations. Countries promote self-sufficiency by setting barriers that limit the import of goods from other places. The approach also restricts local businesses from exporting to other countries.

India: Example of the Self-Sufficiency Approach. For many years India made effective use of many barriers to trade. Businesses were supposed to produce goods for consumption inside India. If private companies were unable to make a profit selling goods only inside India, the government provided subsidies, such as cheap electricity, or wiped out debts. The government owned not just communications, transportation, and power companies, a common feature around the world, but also businesses such as insurance companies and automakers, left to the private sector in most countries.

(303)

Problems with the Self-Sufficiency Alternative. The experience of India and other LDCs revealed two major problems with self-sufficiency: **Inefficiency**: self-sufficiency protects inefficient industries. Companies protected from international competition do not feel pressure to keep abreast of rapid technological changes. **Large bureaucracy**: the second problem . . . was the large bureaucracy needed to administer the controls. A complex administrative system encouraged abuse and corruption.

Development through International Trade
The international trade model of development calls for a country to identify its distinctive or unique economic assets. According to the international trade approach, a country can develop economically by concentrating scarce resources on expansion of its distinctive local industries.

Rostow's Development Model. A pioneering advocate of this approach was W. W. Rostow, who in the 1950s proposed a five-stage model of development. Several countries adopted this approach during the 1960s, although most continued to follow the self-sufficiency approach.

(323)

1. **The traditional society.**
2. **The preconditions for takeoff.**
3. **The takeoff.**
4. **The drive to maturity.**
5. **The age of mass consumption.**

According to the international trade model, each country is in one of these five stages of development. The model assumes that less developed countries will achieve development by moving along from an earlier to a later stage. A country that concentrates on international trade benefits from exposure to consumers in other countries. Concern for international competitiveness in the exporting takeoff industries will filter through less advanced economic sectors. Rostow's optimistic development model was based on two factors. First, the developed countries of Western Europe and Anglo-America had been joined by others in Southern and Eastern Europe and Japan.

(304)
Second, many LDCs contain an abundant supply of raw materials. In the past, European colonial powers extracted many of these resources without paying compensation to the colonies. In a global economy, the sale of these raw materials could generate funds for LDCs to promote development.

Examples of International Trade Approach
Two groups of countries chose the international trade approach during the mid-twentieth century. One such group was along the Arabian Peninsula near the Persian Gulf; the others were in East and Southeast Asia.

Petroleum-Rich Persian Gulf States. This region was one of the world's least developed until the 1970s, when escalation of petroleum prices transformed these countries overnight into some of the wealthiest per capita. Persian Gulf countries have used petroleum revenues to finance large-scale projects. The landscape has been further changed by the diffusion of consumer goods. Some Islamic religious principles, which dominate the culture of the Middle East, conflict with business practices in more developed countries. Women are excluded from holding most jobs and visiting public places.

The Four Asian Dragons. Also among the first countries to adopt the international trade alternative were South Korea, Singapore, Taiwan, and Hong Kong. Singapore and Hong Kong, British colonies until 1965 and 1997, respectively, have virtually no natural resources. Both comprise large cities surrounded by very small amounts of rural land. South Korea and Taiwan have traditionally taken their lead from Japan. Lacking natural resources, the four dragons promoted development by concentrating on producing a handful of manufactured goods.

(324)
Problems with the International Trade Alternative
Three problems have hindered countries outside the Persian Gulf and the four Asian dragons from developing through the international trade approach:
   1. Uneven resource distribution
   2. Market stagnation
   3. Increased dependence on MDCs

Recent Triumph of the International Trade Approach
Despite problems with the international trade approach, it has been embraced by most countries as the preferred alternative for stimulating development. During the past quarter century, world wealth (as measured by GDP) has doubled, whereas world trade has tripled, a measure of the growing importance of the international trade approach. India, for example, dismantled its formidable collection of barriers to international trade during the 1990s.

Countries converted from self-sufficiency to international trade during the 1990s for one simple reason: overwhelming evidence that international trade better promoted development. In the case of

India, under self-sufficiency between 1960 and 1990, GDP grew by 4 percent per year, much lower than in Asian countries that had embraced international trade. After adopting the international trade alternative in the early 1990s, India's GDP grew 7 percent per year during the 1990s.

World Trade Organization. To promote the international trade development model, countries representing 97 percent of world trade established the World Trade Organization (WTO) in 1995. The WTO works to reduce barriers to international trade in two principal ways. First, through the WTO, countries negotiate reduction or elimination of international trade restrictions on manufactured goods and restrictions on the international movement of money by banks, corporations, and wealthy individuals. Second, the WTO promotes international trade by enforcing agreements.

The WTO has been sharply attacked by liberal and conservative critics. Liberal critics charge that the WTO is antidemocratic, because decisions made behind closed doors promote the interest of large corporations rather than the poor. Conservatives charge that the WTO compromises the power and sovereignty of individual countries because it can order changes in taxes and laws that it considers unfair trading practices. Protesters routinely gather in the streets outside high-level meetings of the WTO.

(325)
Financing Development
Regardless of whether self-sufficiency or international trade is preferred, less developed countries lack the money needed to finance development.

Loans. LDCs borrow money to build new infrastructure. The two major lenders are international lending organizations controlled by the MDC governments—the World Bank and the International Monetary Fund. Money is also lent by commercial banks in more developed countries. The theory behind borrowing money to build infrastructure is that . . . new or expanded businesses . . . attracted to an area . . . will contribute additional taxes that the LDC uses in part to repay the loans and in part to improve its citizens' living conditions. The problem is that many of the new infrastructure projects are expensive failures. Many LDCs have been unable to repay the interest on their loans, let alone the principal.
Debt actually exceeds annual income in approximately 30 countries. Financial institutions in more developed countries refuse to make further loans, so construction of needed infrastructure stops. The inability of many LDCs to repay loans also damages the financial stability of banks in the more developed countries.

MDCs have become more cautious in granting loans. International lending agencies require LDCs to adopt **structural adjustment programs**, which are economic policies that create conditions encouraging international trade. These programs can be unpopular with the voters and can encourage political unrest. For their part, LDCs demand an increased role in loan-making decisions made by international agencies.

(326)
Transnational Corporations. A transnational corporation operates in countries other than the one in which its headquarters are located. The net flow of investment from MDCs to LDCs made by private corporations grew nearly tenfold during the 1990s. About one-half of the investment involved transfers within transnational corporations. Foreign investment does not flow equally around the world. In 2001 only one-fourth of foreign investment went from a more developed country to a less developed country.

One-half of all of the international investment in LDCs was clustered in three countries: Brazil, China (including Hong Kong), and Mexico. Another one-fourth was in 20 countries, primarily in South America and Asia.

Key Terms
Development (p.299)
Gender Empowerment Index (GEM) (p.315)
Gender-Related Development Index (GDI) (p.315)
Human Development Index (HDI) (p.299)
Gross domestic product (GDP) (p.300)
Less developed country (LDC) (p.299)
Literacy rate (p.303)
More developed country (MDC) (p.299)
Primary sector (p.300)
Productivity (p.302)
Secondary sector (p.300)
Structural adjustment program (p.325)
Tertiary sector (p.301)
Value added (p.302)

Chapter 10. Agriculture

Providing food in the United States and Canada is a vast industry. The mechanized, highly productive American or Canadian farm contrasts with the subsistence farm found in much of the world. This sharp contrast in agricultural practices constitutes one of the most fundamental differences between the more developed and less developed countries of the world.

Key Issues
1. Where did agriculture originate?
2. Where are agricultural regions in less developed countries?
3. Where are agricultural regions in more developed countries?
4. Why do farmers face economic difficulties?

(333)

The previous chapter divided economic activities into primary, secondary, and tertiary sectors. This chapter is concerned with the principal form of primary-sector economic activity— agriculture. The next two chapters look at the secondary and tertiary sectors. In less developed *regions*, the farm products are most often consumed on or near the farm, whereas in more developed countries farmers sell what they produce. The reason *why* farming varies around the world relates to distribution across *space* of cultural and environmental factors. Elements of the physical environment, such as climate, soil, and topography, set broad limits on agricultural practices, and farmers make choices to modify the environment in a variety of ways. Broad climate patterns influence the crops planted in a region, and local soil conditions influence the crops planted on an individual farm. Farmers choose from a variety of agricultural practices, based on their perception of the value of each alternative. These values are partly economic and partly cultural. How farmers deal with their physical environment varies according to dietary preferences, availability of technology, and other cultural traditions. At a global *scale*, farmers increasingly pursue the most profitable agriculture. After examining the origins and diffusion of agriculture, we will consider the agricultural practices used in less developed and more developed regions.

Key Issue 1. Where Did Agriculture Originate?
- **Origins of agriculture**
- **Location of agricultural hearths**
- **Classifying agricultural regions**

The origins of agriculture cannot be documented with certainty, because it began before recorded history. Scholars try to reconstruct a logical sequence of events based on fragments. Improvements in cultivating plants and domesticating animals evolved over thousands of years.

Origins of Agriculture
Determining the origin of agriculture first requires a definition of what it is—and agriculture is not easily defined. We will use this definition: **Agriculture** is deliberate modification of Earth's surface through cultivation of plants and rearing of animals to obtain sustenance or economic gain.

Hunters and Gatherers
Before the invention of agriculture, all humans probably obtained the food they needed for survival through hunting for animals, fishing, or gathering. Hunters and gatherers lived in small groups. The men hunted game or fished, and the women collected berries, nuts, and roots. This division of labor sounds like a stereotype but is based on evidence from archaeology and anthropology. The group traveled frequently, establishing new home bases or camps. The direction and frequency of migration depended on the movement of game and the seasonal growth of plants at various locations.

(334)

Contemporary Hunting and Gathering. Today perhaps a quarter-million people, or less than 0.005 percent of the world's population, still survive by hunting and gathering. Contemporary hunting and gathering societies are isolated groups living on the periphery of world settlement, but they provide insight into human customs that prevailed in prehistoric times, before the invention of agriculture.

Invention of Agriculture
Why did nomadic groups convert from hunting, gathering, and fishing to agriculture? Over thousands of years, plant cultivation apparently evolved from a combination of accident and deliberate experiment.

Two Types of Cultivation. The earliest form of plant cultivation, according to . . . Carl Sauer, was **vegetative planting**, direct cloning from existing plants, such as cutting stems and dividing roots. Coming later, according to Sauer, was **seed agriculture**. Seed agriculture is practiced by most farmers today.

Location of Agricultural Hearths
Agriculture probably did not originate in one location, but began in multiple, independent hearths.

Location of First Vegetative Planting
Sauer believes that vegetative planting probably originated in Southeast Asia. The region's diversity of climate and topography . . . encouraged . . . plants suitable for dividing. Also, the people obtained food primarily by fishing rather than by hunting and gathering, so they may have been more sedentary and therefore able to devote more attention to growing plants. The first plants domesticated in Southeast Asia . . . probably included roots such as the taro and yam, and tree crops such as the banana and palm. The dog, pig, and chicken probably were domesticated first in Southeast Asia. Other early hearths of vegetative planting also may have emerged independently in West Africa and northwestern South America.

Location of First Seed Agriculture
Seed agriculture also originated in more than one hearth. Sauer identified three hearths in the Eastern Hemisphere: western India, northern China, and Ethiopia. Seed agriculture diffused quickly from western India to Southwest Asia, where important early advances were made, including the domestication of wheat and barley.

(335)
Apparently, inhabitants of Southwest Asia also were first to integrate seed agriculture with domestication of herd animals such as cattle, sheep, and goats. This integration of plants and animals is a fundamental element of modern agriculture.

Diffusion of Seed Agriculture. Seed agriculture diffused from Southwest Asia across Europe and through North Africa. Greece, Crete, and Cyprus display the earliest evidence of seed agriculture in Europe.
Seed agriculture also diffused eastward from Southwest Asia to northwestern India and the Indus River plain. Again, various domesticated plants and animals were brought from Southwest Asia, although other plants, such as cotton and rice, arrived in India from different hearths. From the northern China hearth, millet diffused to South Asia and Southeast Asia. Rice . . . has an unknown hearth. Sauer identified a third independent hearth in Ethiopia, where millet and sorghum were domesticated early. However, he argued that agricultural advances in Ethiopia did not diffuse widely to other locations.

Two independent seed agriculture hearths originated in the Western Hemisphere: southern Mexico and northern Peru. Agricultural practices diffused to other parts of the Western Hemisphere. That agriculture had multiple origins means that, from earliest times, people have produced food in distinctive ways in different regions. This diversity derives from a unique legacy of wild plants, climatic conditions, and cultural preferences in each region. Improved communications in recent centuries have encouraged the diffusion of some plants to varied locations around the world.

Classifying Agricultural Regions
The most fundamental differences in agricultural practices are between those in less developed countries and those in more developed countries.

Differences between Subsistence and Commercial Agriculture
Subsistence agriculture . . . is the production of food primarily for consumption by the farmer's family. **Commercial agriculture . . .** is the production of food primarily for sale off the farm. Five principal features distinguish commercial . . . from subsistence agriculture: purpose of farming; percentage of farmers in the labor force; use of machinery; farm size; (and) relationship of farming to other businesses.

(336)
Purpose of Farming. In LDCs most people produce food for their own consumption. Some surplus may be sold . . . but . . . may not even exist some years. In commercial farming, farmers grow crops and raise animals primarily for sale. Agricultural products are . . . sold . . . to food-processing companies.

Percentage of Farmers in the Labor Force. In more developed countries less than 5 percent of the workers are engaged directly in farming, compared to 55 percent in less developed countries. The percentage of farmers is even lower in the United States and Canada, at only 2 percent. The number of farmers has declined dramatically in more developed societies during the twentieth century. Both push and pull migration factors have been responsible.

Use of Machinery. A small number of farmers in more developed societies can feed many people because they rely on machinery to perform work. Traditionally, the farmer or local craftspeople made equipment from wood, but beginning in the late eighteenth century, factories produced farm machinery. The first all-iron plow was made in the 1770s. Factory-made farm machines have replaced or supplemented manual labor.

Transportation improvements also aid commercial farmers. Railroads in the nineteenth century, and highways and trucks in the twentieth century, have enabled farmers to transport crops and livestock farther and faster. Commercial farmers use scientific advances to increase productivity.

(337)
Some farmers conduct their own on-farm research. Electronics also aid commercial farmers. Global Positioning Systems units determine precise coordinates for spreading different types and amounts of fertilizers. Both satellite imagery and yield monitors attached to combines monitor production yields.

Farm Size. The average farm size is relatively large in commercial agriculture, especially in the United States and Canada. Commercial agriculture is increasingly dominated by a handful of large farms. In the United States the largest 4 percent of farms . . . account for more than one half of the country's total output. One half of U.S. farms generate less than $10,000 a year in sales. Large size is partly a consequence of mechanization. As a result of the large size and the high level of mechanization, commercial agriculture is an expensive business.

(338)
This money is frequently borrowed from a bank and repaid after the output is sold. Although the United States currently has fewer farms and farmers than in 1900, the amount of land devoted to agriculture has increased. However, the amount of U.S. farmland has declined from an all-time peak around 1960. A . . . serious problem in the United States has been the loss of the most productive farmland, known as **prime agricultural land**, as urban areas sprawl into the surrounding countryside.

Relationship of Farming to Other Businesses. Commercial farming is closely tied to other businesses. Commercial farming . . . has been called **agribusiness**, . . . integrated into a large food-production industry.

(339)
Although farmers are less than 2 percent of the U.S. labor force, more than 20 percent of U.S. labor works in food production related to agribusiness: food processing, packaging, storing, distributing, and retailing.

Mapping Agricultural Regions

Several attempts have been made to outline the major types of subsistence and commercial agriculture currently practiced in the world, but few of these classifications include maps that show regional distributions. The most widely used map of world agricultural regions was prepared by geographer Derwent Whittlesey in 1936. Whittlesey identified 11 main agricultural regions, plus an area where agriculture was nonexistent. Whittlesey sorted out agricultural practices primarily by climate. Agriculture varies between the drylands and the tropics within LDCs—as well as between the drylands of less developed and more developed countries.

Because of the problems with environmental determinism discussed in Chapter 1, geographers are wary of placing too much emphasis on the role of climate. Cultural preferences, discussed in Chapter 4, explain some agricultural differences in areas of similar climate.

Key Issue 2. Where Are Agricultural Regions in Less Developed Countries?
- **Shifting cultivation**
- **Pastoral nomadism**
- **Intensive subsistence agriculture**

This section considers four of the five agricultural types characteristic of LDCs: shifting cultivation, pastoral nomadism, and two types of intensive subsistence. The fifth type of LDC agriculture, plantation, is discussed in the third key issue along with agriculture in more developed countries.

Shifting Cultivation

Shifting cultivation is practiced in much of the world's Humid Low-Latitude, or A, climate regions, which have relatively high temperatures and abundant rainfall. It is called shifting cultivation rather than shifting agriculture because "agriculture" implies greater use of tools and animals and more sophisticated modification of the landscape.

Characteristics of Shifting Cultivation

Shifting cultivation has two distinguishing hallmarks: farmers clear land for planting by slashing vegetation and burning the debris; and farmers grow crops on a cleared field for only a few years. People who practice shifting cultivation generally live in small villages and grow food on the surrounding land, which the village controls.

(341)
The Process of Shifting Cultivation. Each year villagers designate (an area) for planting. They must remove the dense vegetation that typically covers tropical land. Using axes, they cut most of the trees, sparing only those that are economically useful. The debris is burned under carefully controlled conditions. Rains wash the fresh ashes into the soil, providing needed nutrients. The cleared area is known by a variety of names in different regions, including **swidden**, *ladang*, *milpa*, *chena*, and *kaingin*. The cleared land can support crops only briefly, usually three years or less. Villagers . . . leave the old site uncropped for many years. The villagers will return to the site, . . . perhaps as few as 6 years or as many as 20 years later, to begin the process of clearing the land again. In the meantime, they may still care for fruit-bearing trees on the site.

(342)
Crops of Shifting Cultivation. The precise crops grown by each village vary by local custom and taste. The predominant crops include upland rice in Southeast Asia, maize (corn) and manioc (cassava) in South America, and millet and sorghum in Africa. Yams, sugarcane, plantain, and

vegetables also are grown in some regions. The Kayapo people of Brazil's Amazon tropical rain forest . . . plant in concentric rings. Plants that require more nutrients are located in the outer ring. It is here that the leafy crowns of cut trees fall when the field is cleared. Most families grow only for their own needs, so one swidden may contain a large variety of intermingled crops. Families may specialize in a few crops and trade with villagers who have a surplus of others.

Ownership and Use of Land in Shifting Cultivation. Traditionally, land is owned by the village as a whole rather than separately by each resident. Private individuals now own the land in some communities, especially in Latin America. Shifting cultivation occupies approximately one fourth of the world's land area, a higher percentage than any other type of agriculture. However, only 5 percent of the world's population engages in shifting cultivation.

Future of Shifting Cultivation
The percentage of land devoted to shifting cultivation is declining in the tropics at the rate of about 100,000 square kilometers (40,000 square miles), or 1 percent per year. The amount of Earth's surface allocated to tropical rain forests has already been reduced to less than half of its original area. Shifting cultivation is being replaced by logging, cattle ranching, and cultivation of cash crops. Until recent years the World Bank supported deforestation with loans to finance development schemes that required clearing forests.

(343)
To its critics, shifting cultivation is at best a preliminary step in economic development. Critics say it then should be replaced by more sophisticated agriculture that yields more per land area.

But defenders of shifting cultivation consider it the most environmentally sound approach for the tropics. Practices used in other forms of agriculture . . . may damage the soil, cause severe erosion, and upset balanced ecosystems. Large-scale destruction of the rain forests also may contribute to global warming. When large numbers of trees are cut, their burning and decay release large volumes of carbon dioxide. Elimination of shifting cultivation could also upset the traditional local diversity of cultures in the tropics. The activities of shifting cultivation are intertwined with other social, religious, political, and various folk customs.

As the importance of tropical rain forests to the global environment has become recognized, LDCs have been pressured to restrict further destruction of them. In Brazil's Amazon rain forest, deforestation is increasing. A 1997 U.S. government study placed deforestation . . . at 58,000 square kilometers (22,000 square miles) per year.

Pastoral Nomadism
Pastoral nomadism is a form of subsistence agriculture based on the herding of domesticated animals. The word *pastoral* refers to sheep herding. It is adapted to dry climates, where planting crops is impossible. Only about 15 million people are pastoral nomads, but they sparsely occupy about 20 percent of Earth's land area.

Characteristics of Pastoral Nomadism
Pastoral nomads depend primarily on animals rather than crops for survival. The animals provide milk, and their skins and hair are used for clothing and tents. Like other subsistence farmers, though, pastoral nomads consume mostly grain rather than meat. Some pastoral nomads obtain grain from sedentary subsistence farmers in exchange for animal products. More often, part of a nomadic group—perhaps the women and children—may plant crops at a fixed location while the rest of the group wanders with the herd. Other nomads might sow grain in recently flooded areas and return later in the year to harvest the crop.

Choice of Animals. Nomads select the type and number of animals for the herd according to local cultural and physical characteristics. The choice depends on the relative prestige of animals and the ability of species to adapt to a particular climate and vegetation.

Movements of Pastoral Nomads. Pastoral nomads do not wander randomly across the landscape but have a strong sense of territoriality. Every group controls a piece of territory and will invade another group's territory only in an emergency or if war is declared. The precise migration patterns evolve from intimate knowledge of the area's physical and cultural characteristics. The selection of routes varies in unusually wet or dry years and is influenced by the condition of their animals and the area's political stability. Some pastoral nomads practice **transhumance**, which is seasonal migration of livestock between mountains and lowland pasture areas.

The Future of Pastoral Nomadism

Agricultural experts once regarded pastoral nomadism as a stage in the evolution of agriculture. Because they had domesticated animals but not plants, pastoral nomads were considered more advanced than hunters and gatherers but less advanced than settled farmers. Pastoral nomadism is now generally recognized as an offshoot of sedentary agriculture, not as a primitive precursor of it. It is simply a practical way of surviving on land that receives too little rain for cultivation of crops. Today pastoral nomadism is a declining form of agriculture, partly a victim of modern technology. Nomads used to be the most powerful inhabitants of the drylands, but now, with modern weapons, national governments can control the nomadic population more effectively. Government efforts to resettle nomads have been particularly vigorous in China, Kazakhstan, and several Middle Eastern countries, including Egypt, Israel, Saudi Arabia, and Syria. Governments force groups to give up pastoral nomadism because they want the land for other uses. In the future, pastoral nomadism will be increasingly confined to areas that cannot be irrigated or that lack valuable raw materials.

Intensive Subsistence Agriculture

Shifting cultivation and pastoral nomadism are . . . found in regions of low (population) density.

But three-fourths of the world's people live in LDCs, and another form of subsistence agriculture is needed to feed most of them: **intensive subsistence agriculture**. In densely populated East, South, and Southeast Asia, most farmers practice intensive subsistence agriculture. The typical farm . . . is much smaller than elsewhere in the world. Because the agricultural density—the ratio of farmers to arable land—is so high in parts of East and South Asia, families must produce enough food for their survival from a very small area of land. They do this through careful agricultural practices, refined over thousands of years in response to local environmental and cultural patterns. Intensive subsistence farmers waste virtually no land. Paths and roads are kept as narrow as possible to minimize the loss of arable land. Little grain is grown to feed the animals.

Intensive Subsistence with Wet Rice Dominant

Wet rice occupies a relatively small percentage of Asia's agricultural land but is the region's most important source of food. Intensive wet-rice farming is the dominant type of agriculture in Southeast China, East India, and much of Southeast Asia.

Successful production of large yields of rice is an elaborate process, time-consuming and done mostly by hand. Growing rice involves several steps: First, a farmer prepares the field for planting, using a plow drawn by water buffalo or oxen. The use of a plow and animal power is one characteristic that distinguishes subsistence agriculture from shifting cultivation. Then the plowed land is flooded with water . . . from rainfall, river overflow, or irrigation. The flooded field is called a **sawah** in the Austronesian language widely spoken in Indonesia, including Java. Europeans and North Americans frequently, but incorrectly, call it a **paddy**, the Malay word for wet rice. Wet rice is most easily grown on flat land, because the plants are submerged in water much of the time. One method of developing additional land suitable for growing rice is to terrace the hillsides of river valleys.

Land is used even more intensively in parts of Asia by obtaining two harvests per year from one field, a process known as **double cropping**. Double cropping is common in places having warm winters . . . but is relatively rare in India, where most areas have dry winters. Normally, double cropping involves alternating between wet rice . . . and wheat, barley, or another dry crop, grown in the drier winter season.

Intensive Subsistence with Wet Rice Not Dominant
Climate prevents . . . growing wet rice in portions of Asia, especially where summer precipitation levels are too low and winters are too harsh. This region shares most of the characteristics of intensive subsistence agriculture with the wet-rice region. Wheat is the most important crop, followed by barley. Other grains and legumes are grown for household consumption (and) . . . some crops sold for cash, such as cotton, flax, hemp, and tobacco.

(346)
In milder parts of the region, more than one harvest can be obtained some years through skilled use of **crop rotation**.

Since the (Chinese) Communist Revolution in 1949, . . . the . . . government organized agricultural producer communes. By combining several small fields into a single large unit, the government hoped to promote agricultural efficiency.

(347)
China has dismantled the agricultural communes. The communes still hold legal title to agricultural land, but villagers sign contracts entitling them to farm portions of the land as private individuals. Reorganization has been difficult because . . . infrastructure was developed to serve large communal farms rather than small, individually managed ones.

Key Issue 3. Where Are Agricultural Regions in More Developed Countries?
- **Mixed crop and livestock farming**
- **Dairy farming**
- **Grain farming**
- **Livestock ranching**
- **Mediterranean agriculture**
- **Commercial gardening and fruit farming**
- **Plantation agriculture**

Commercial agriculture in more developed countries can be divided into six main types. Each type is predominant in distinctive regions within MDCs, depending largely on climate. The end of this section examines plantation farming, a form of commercial agriculture in LDCs.

Mixed Crop and Livestock Farming
Mixed crop and livestock farming is the most common form of commercial agriculture in the United States west of the Appalachians and east of 98° west longitude and in much of Europe from France to Russia.

Characteristics of Mixed Crop and Livestock Farming
The most distinctive characteristic of mixed crop and livestock farming is its integration of crops and livestock. Most of the crops are fed to animals rather than consumed directly by humans. Mixed crop and livestock farming permits farmers to distribute the workload more evenly through the year . . . (and) reduces seasonal variations in income.

Crop Rotation Systems. Mixed crop and livestock farming typically involves crop rotation. Crop rotation contrasts with shifting cultivation, in which nutrients depleted from a field are restored only by leaving the field fallow (uncropped) for many years. A two-field crop-rotation system was developed in Northern Europe as early as the fifth century A.D. Beginning in the eighth century, a three-field system was introduced. Each field yielded four harvests every six years, compared to three every six years under the two-field system. A four-field system was used in Northwest Europe by the eighteenth century.

(348)
Each field thus passed through a cycle of four crops: root, cereal, rest crop, and another cereal. Cereals . . . were sold for flour and beer production, and straw . . . was retained for animal bedding. Root crops . . . were fed to the animals during the winter. Clover and other "rest" crops were used for cattle grazing and restoration of nitrogen to the soil.

Choice of Crops
In the United States, mixed crop and livestock farmers select corn most frequently because of higher yields per area than other crops. Some of the corn is consumed by people . . . but most is fed to pigs and cattle. Soybeans have become the second most important crop in the U.S. mixed commercial farming region . . . mostly to make animal feed.

Dairy Farming
Dairy farming is the most important type of commercial agriculture practiced on farms near the large urban areas of the Northeast United States, Southeast Canada, and Northwest Europe. Russia, Australia, and New Zealand also have extensive areas devoted to dairy farming. Nearly 60 percent of the world's supply of milk is produced and consumed in these developed regions. Traditionally, fresh milk was rarely consumed except directly on the farm or in nearby villages. During the nineteenth century, demand for the sale of milk to urban residents increased. Rising incomes permitted urban residents to buy milk products, which were once considered luxuries.

Why Dairy Farms Locate Near Urban Areas
Dairying has become the most important type of commercial agriculture in the first ring outside large cities because of transportation factors.

(349)
The ring surrounding a city from which milk can be supplied without spoiling is known as the **milkshed**. Improvements in transportation have permitted dairying to be undertaken farther from the market. As a result, nearly every farm in the U.S. Northeast and Northwest Europe is within the milkshed of at least one urban area.

Some dairy farms specialize in products other than milk. Originally, butter and cheese were made directly on the farm, primarily from the excess milk produced in the summer, before modern agricultural methods evened the flow of milk through the year.

Regional Differences in Dairy Products
The choice of product varies within the U.S. dairy region, depending on whether the farms are within the milkshed of a large urban area. Farms located farther from consumers are more likely to sell their output to processors. In the East, virtually all milk is sold to consumers living in . . . large urban areas. Farther west, most milk is processed into cheese and butter. Countries likewise tend to specialize in certain products. New Zealand, the world's largest producer of dairy products, devotes about 5 percent to liquid milk, compared to over 50 percent in the United Kingdom. Dairy farmers, like other commercial farmers, usually do not sell their products directly to consumers.

Problems for Dairy Farmers
Like other commercial farmers, dairy farmers face economic problems because of declining revenues and rising costs. Dairy farming is labor-intensive.

(350)
Dairy farmers also face the expense of feeding the cows in the winter, when they may be unable to graze on grass. The number of farms with milk cows declined in the United States by two-thirds

between 1980 and 2000. The number of dairy cows declined by only one-eighth, and production actually increased by one-fourth—yields per cow increased substantially.

Grain Farming
Commercial grain agriculture is distinguished from mixed crop and livestock farming because crops on a grain farm are grown primarily for consumption by humans rather than by livestock. Wheat generally can be sold for a higher price than other grains . . . and it has more uses as human food. Because wheat has a relatively high value per unit weight, it can be shipped profitably from remote farms to markets.

Grain-farming Regions
The United States is by far the largest commercial producer of grain. Large-scale commercial grain production is found in only a few other countries, including Canada, Argentina, Australia, France, and the United Kingdom. Commercial grain farms are generally located in regions that are too dry for mixed crop and livestock agriculture.

Within North America, large-scale grain production is concentrated in three areas. The first is the **winter wheat** belt that extends through Kansas, Colorado, and Oklahoma. The second important grain-producing region in North America is the **spring wheat** belt of the Dakotas, Montana, and southern Saskatchewan in Canada. A third important grain-growing region is the Palouse region of Washington State.

Large-scale grain production, like other commercial farming ventures in more developed countries, is heavily mechanized, conducted on large farms, and oriented to consumer preferences. Unlike work on a mixed crop and livestock farm, the effort required to grow wheat is not uniform throughout the year. Some individuals or firms may therefore have two sets of fields—one in the spring-wheat belt and one in the winter-wheat belt.

(351)
The same machinery can be used in the two regions, thus spreading the cost of the expensive equipment. Combine companies start working in Oklahoma in early summer and work their way northward.

Importance of Wheat
Wheat is grown to a considerable extent for international trade and is the world's leading export crop. The ability to provide food for many people elsewhere in the world is a major source of economic and political strength for the United States and Canada.

Livestock Ranching.
Ranching is the commercial grazing of livestock over an extensive area, . . . practiced in more developed countries, where the vegetation is too sparse and the soil too poor to support crops.

Cattle Ranching in U.S. Popular Culture
The importance of ranching in the United States extends beyond the number of people who choose this form of commercial farming because of its prominence in popular culture. Cattle ranching in Texas, though, as glamorized in popular culture, actually dominated commercial agriculture for a short period—from 1867 to 1885.

Beginning of U.S. Cattle Ranching. Cattle were first brought to the Americas by Columbus on his second voyage. Living in the wild, the cattle multiplied and thrived on abundant grazing lands on the frontiers of North and South America. Immigrants from Spain and Portugal—the only European countries with a tradition of cattle ranching—began ranching in the Americas. Cattle ranching in the United States expanded because of demand for beef in the East Coast cities during the 1860s. Ranchers who could get their cattle to Chicago were paid $30 to $40 per head, compared to only $3 or $4 per head in Texas.

(352)

Transporting Cattle to Market. To reach Chicago, cattle were driven on hoof by cowboys over trails from Texas to the nearest railhead. The western terminus of the rail line reached Abilene, Kansas, in 1867. The number of cattle brought into Abilene increased from 1,000 in 1867 to 35,000 in 1868 and 150,000 in 1869. After a few years the terminus of the railroad moved farther west. The most famous route from Texas northward to the rail line was the Chisholm Trail. Today U.S. Route 81 roughly follows the course of the Chisholm Trail.

Fixed Location Ranching

Cattle ranching declined in importance during the 1880s after it came in conflict with sedentary agriculture. The early cattle ranchers in the West owned little land, only cattle.

(353)

Range Wars. The U.S. government, which owned most of the land used for open grazing, began to sell it to farmers to grow crops. For a few years the ranchers tried to drive out the farmers. The farmers' most potent weapon proved to be barbed wire, first commercially produced in 1873. Ranchers were compelled to buy or lease land to accommodate their cattle. Sixty percent of cattle grazing today are on land leased from the U.S. government.

Changes in Cattle Breeding. Ranchers were also induced to switch from cattle drives to fixed-location ranching by a change in the predominant breed of cattle. Longhorns . . . were hardy animals . . . but . . . the meat of longhorns was of poor quality. New cattle breeds introduced from Europe, such as the Hereford, offered superior meat but were not adapted to the old ranching system. These breeds thrived once open grazing was replaced by fixed ranching, and long-distance trail drives and rail journeys to Chicago gave way to short rail or truck trips to nearby meat packers. With the spread of irrigation techniques and hardier crops, land in the United States has been converted from ranching to crop growing. Cattle are still raised on ranches but are frequently sent for fattening to farms or to local feed lots.

Ranching outside the United States

Commercial ranching is conducted in other more developed regions of the world.

(354)

Ranching is rare in Europe, except in Spain and Portugal. In South America a large portion of the pampas . . . are devoted to grazing cattle and sheep. The relatively humid climate on the pampas provides more shoots and shrubs on a given area of land than in the U.S. West. Land was divided into large holdings in the nineteenth century, in contrast to the U.S. practice. Ranching has declined in Argentina . . . because growing crops is more profitable except on very dry lands. The interior of Australia was opened for grazing in the nineteenth century, although sheep are more common than cattle. Ranches in the Middle East, New Zealand, and South Africa are also more likely to have sheep.

Ranching has followed similar stages around the world: first, . . . herding . . . over open ranges, then . . . . transformed into fixed farming by dividing the open land. Many of the farms converted to growing crops, and ranching was confined to the drier lands. Ranching (later) became part of the meat-processing industry rather than an economic activity carried out on isolated farms.

Mediterranean Agriculture

Mediterranean agriculture exists primarily in the lands that border the Mediterranean Sea. Farmers in California, central Chile, the southwestern part of South Africa, and southwestern Australia practice Mediterranean agriculture as well. Every Mediterranean area borders a sea. Prevailing sea winds provide moisture and moderate the winter temperatures. Summers are hot and dry. The land is very hilly. Farmers derive a smaller percentage of income from animal products in the Mediterranean region than in the mixed crop and livestock region. Some farmers living along the Mediterranean Sea traditionally used transhumance to raise animals, although the practice is now less common.

Mediterranean Crops
Most crops in Mediterranean lands are grown for human consumption rather than for animal feed. **Horticulture**—which is the growing of fruits, vegetables, and flowers—and tree crops form the commercial base of the Mediterranean farming. A combination of local physical and cultural characteristics determines which crops are grown in each area. In the lands bordering the Mediterranean Sea, the two most important cash crops are olives and grapes.

Despite the importance of olives and grapes to commercial farms bordering the Mediterranean Sea, approximately half of the land is devoted to growing cereals, especially wheat for pasta and bread.

(355)
Cereals occupy a much lower percentage of the cultivated land in California than in other Mediterranean climates. Instead, much of California farmland is devoted to fruit and vegetable horticulture. The rapid growth of urban areas in California, especially Los Angeles, has converted high-quality agricultural land into housing developments. The loss of farmland has been offset by expansion of agriculture into arid lands. However, farming in drylands requires massive irrigation to provide water.

Commercial Gardening and Fruit Farming
Commercial gardening and fruit farming is the predominant type of agriculture in the U.S. Southeast, . . . frequently called **truck farming**, because "truck" was a Middle English word meaning bartering or the exchange of commodities. Truck farms grow . . . fruits and vegetables. Some of these fruits and vegetables are sold fresh to consumers, but most are sold to large processors. Truck farms are highly efficient large-scale operations that take full advantage of machines at every stage of the growing process. Labor costs are kept down by hiring migrant farm workers. A handful of farms may dominate national output of some fruits and vegetables. A form of truck farming called *specialty farming* has spread to New England, . . . growing crops that have limited but increasing demand among affluent consumers.

Plantation Farming
The plantation is a form of commercial agriculture found in the tropics and subtropics, especially in Latin America, Africa, and Asia. Plantations are often owned or operated by Europeans or North Americans and grow crops for sale primarily in more developed countries. A **plantation** is a large farm that specializes in one or two crops. Among the most important crops . . . are cotton, sugarcane, coffee, rubber, and tobacco, . . . cocoa, jute, bananas, tea, coconuts, and palm oil.

Because plantations are usually situated in sparsely settled locations, they must import workers. Managers try to spread the work . . . throughout the year to make full use of the large labor force. Crops such as tobacco, cotton, and sugarcane, which can be planted only once a year, are less likely to be grown on large plantations today than in the past. Crops are normally processed at the plantation. Processed goods are less bulky and therefore cheaper to ship. Until the Civil War, plantations were important in the U.S. South, where the principal crop was cotton, followed by tobacco and sugarcane. Slaves brought from Africa performed most of the labor until . . . the defeat of the South in the Civil War. Thereafter, plantations . . . were subdivided and either sold to individual farmers or worked by tenant farmers.

Key Issue 4. Why Do Farmers Face Economic Difficulties?
- **Issues for commercial farmers**
- **Issues for subsistence farmers**
- **Strategies to increase food supply**

(356)
Issues for Commercial Farmers
Two economic factors influence the choice of crops (or livestock) by commercial farmers: access to markets and overproduction.

Access to Markets
Because the purpose of commercial farming is to sell produce off the farm, the distance from the farm to the market influences the farmer's choice of crop to plant. Geographers use the von Thünen model to help explain the importance of proximity to market in the choice of crops on commercial farms.

Von Thünen's Model. The von Thünen model was first proposed in 1826 by Johann Heinrich von Thünen, a farmer in northern Germany, in a book titled *The Isolated State*. In choosing an enterprise, a commercial farmer compares two costs: the cost of the land versus the cost of transporting products to market. First, a farmer identifies a crop that can be sold for more than the land cost. A farmer will not necessarily plant the crop that sells for the highest price per hectare. Distance to market is critical because the cost of transporting each product is different.

Example of Von Thünen's Model. The example shows that a farmer would make a profit growing wheat on land located less than 4 kilometers from the market. Beyond 4 kilometers, wheat is not profitable, because the cost of transporting it exceeds the gross profit. More distant farms are more likely to select crops that can be transported less expensively.

Application of Von Thünen's Model. Von Thünen based his general model of the spatial arrangement of different crops on his experiences as owner of a large estate in northern Germany during the early nineteenth century. He found that specific crops were grown in different rings around the cities in the area.
Von Thünen did not consider site or human factors in his model, . . . although he recognized that the model could vary according to topography and other distinctive physical conditions. The model also failed to understand that social customs and government policies influence the attractiveness of plants and animals for a commercial farmer. Although von Thünen developed the model for a small region with a single market center, it also applies to a national or global scale.

(357)
Overproduction in Commercial Farming
Commercial farmers suffer from low incomes because they produce too much food rather than too little. A surplus of food has been produced in part because of widespread adoption of efficient agricultural practices. Commercial farmers have dramatically increased the capacity of the land to produce food. While the food supply has increased in more developed countries, demand has remained constant, because the market for most products is already saturated. Demand is also stagnant for most agricultural products in more developed countries because of low population growth.

U.S. Government Policies. The U.S. government has three policies to attack the problem of excess productive capacity. First, farmers are encouraged to avoid producing crops that are in excess supply. The government encourages planting fallow crops. Second, the government pays farmers when certain commodity prices are low. Third, the government buys surplus production and sells or donates it to foreign governments. In addition, low-income Americans receive food stamps in part to stimulate their purchase of additional food. The United States spends about $10 billion a year on farm subsidies. Annual spending varies considerably from one year to the next. Government policies point out a fundamental irony in worldwide agricultural patterns. In a more developed country such as the United States, farmers are encouraged to grow less food, while less developed countries struggle to increase food production to match the rate of the growth in population.

Sustainable Agriculture
Some commercial farmers are converting their operations to sustainable agriculture, an agricultural practice that preserves and enhances environmental quality. Farmers practicing sustainable agriculture typically generate lower revenues than do conventional farmers, but they also have lower costs. Two principal practices distinguish sustainable agriculture from conventional agriculture:
  1. More sensitive land management
  2. Better integration of crops and livestock

Sensitive Land Management. Sustainable agriculture protects soil in part through ridge tillage and limited use of chemicals. **Ridge tillage** is a system of planting crops on 4-to 8-inch ridges that are formed during cultivation or after harvest. Ridge tillage is attractive for two main reasons: lower production costs and greater soil conservation. Production costs are lower with ridge tillage in part because it requires less investment in tractors and other machinery than conventional planting.

(358)

Ridge tillage features a minimum of soil disturbance from harvest to the next planting. Over several years the soil will tend to have increased organic matter, greater water holding capacity and more earthworms. The channels left by earthworms and decaying roots enhance drainage. Under sustainable agriculture, farmers control weeds with cultivation and minimal use of herbicides.

Integrated Crop and Livestock. Sustainable agriculture attempts to integrate the growing of crops and the raising of livestock as much as possible at the level of the individual farm. Animals consume crops grown on the farm and are not confined to small pens.

Issues for Subsistence Farmers

Two economic issues discussed in earlier chapters influence the choice of crops planted by subsistence farmers: first, . . . rapid population growth, (and) second, . . . adopting the international trade approach to development.

Subsistence Farming and Population Growth

According to Ester Boserup, population growth compels subsistence farmers to consider new farming. For hundreds if not thousands of years, subsistence farming . . . yielded enough food. Suddenly in the late twentieth century, the LDCs needed to provide enough food for a rapidly increasing population.

According to the Boserup thesis, subsistence farmers increase the supply of food through intensification of production, achieved in two ways. First, land is left fallow for shorter periods. Boserup identified five basic stages in the intensification of farmland: Forest Fallow; Bush Fallow; Short Fallow; Annual Cropping; and Multicropping. Eventually, farmers achieve the very intensive use of farmland characteristic of areas of high population density. The second way that subsistence farmers intensify production, according to the Boserup thesis, is through adopting new farming methods. The additional labor needed to perform these operations comes from the population growth.

(359)

Subsistence Farming and International Trade

To expand production, subsistence farmers need higher-yield seeds, fertilizer, pesticides, and machinery. For many African and Asian countries . . . the main source of agricultural supplies is importing. To generate the funds they need to buy agricultural supplies, less developed countries must produce something they can sell in more developed countries. In a less developed country such as Kenya, families may divide by gender between traditional subsistence agriculture and contributing to international trade. The more land that is devoted to growing export crops, the less that is available to grow crops for domestic consumption. Rather than helping to increase productivity, the funds generated through the sale of export crops may be needed to feed the people who switched from subsistence farming to growing export crops.

Drug Crops. The export crops chosen in some LDCs, especially in Latin America and Asia, are those that can be converted to drugs. Various drugs, such as coca leaf, marijuana, opium, and hashish, have distinctive geographic distributions.

Strategies to Increase Food Supply

Four strategies can increase the food supply:
1.   Expand the land area used for agriculture
2.   Increase the productivity of land now used for agriculture
3.   Identify new food sources
4.   Increase exports from other countries

Increase Food Supply by Expanding Agricultural Land. Historically, world food production increased primarily by expanding the amount of land devoted to agriculture. Today few scientists believe that further expansion of agricultural land can feed the growing world population. Beginning about 1950, the human population has increased faster than the expansion of agricultural land. Prospects for expanding the percentage of cultivated land are poor in much of Europe, Asia, and Africa.

Especially in semiarid regions, human actions are causing land to deteriorate to a desertlike condition, a process called desertification (more precisely, semiarid land degradation). The United Nations estimates that desertification removes 27 million hectares (104,000 square miles) of land from agricultural production each year, an area roughly equivalent to Colorado.

Excessive water threatens other agricultural areas, especially drier lands that receive water from human-built irrigation systems. The United Nations estimates that 10 percent of all irrigated land is waterlogged, mostly in Asia and South America.

(360)
As urban areas grow in population and land area, farms on the periphery are replaced by homes, roads, shops, and other urban land uses.

Increase Food Supply through Higher Productivity. The invention and rapid diffusion of more productive agricultural techniques during the 1970s and 1980s is called the green revolution. The green revolution involves two main practices: the introduction of new higher-yield seeds and the expanded use of fertilizers.

The new high yield wheat, rice and maize seeds were diffused rapidly around the world. India's wheat production, for example, more than doubled in five years. Other Asian and Latin American countries recorded similar productivity increases.

To take full advantage of the new miracle seeds, farmers must use more fertilizer and machinery. The problem is that the cheapest way to produce both types of nitrogen-based fertilizers is to obtain hydrogen from natural gas or petroleum. As fossil-fuel prices increase, so do the prices for nitrogen-based fertilizers, which then become too expensive for many farmers in LDCs.

(361)
Farmers need tractors, irrigation pumps, and other machinery to make the most effective use of the new miracle seeds. In LDCs, farmers cannot afford such equipment, nor, in view of high energy costs, can they buy fuel to operate the equipment.

Scientists have continued to create higher-yield hybrids that are adapted to environmental conditions in specific regions. The green revolution was largely responsible for preventing a food crisis in these regions during the 1970s and 1980s, but will these scientific breakthroughs continue in the twenty-first century?

Increase Food Supply by Identifying New Food Sources. The third alternative for increasing the world's food supply is to develop new food sources. Three strategies being considered are to cultivate the oceans, to develop higher-protein cereals, and to improve palatability of rarely consumed foods.

(362)
Increase Food Supply by Increasing Exports from Other Countries. The fourth alternative for increasing the world's food supply is to export more food from countries that produce surpluses. The three top export grains are wheat, maize (corn), and rice. Few countries are major exporters of food, but increased production in these countries could cover the gap elsewhere.

The United States remains by far the largest grain exporter, accounting for one-half of global corn exports and one-fourth of wheat. However, the United States has decreased its grain exports in the past quarter century, whereas other countries have increased theirs.

(363)
Japan is by far the world's leading grain importer, especially of corn and wheat. South Korea and Mexico are major importers of corn, Egypt and Italy of wheat. World volume of trade in rice is much lower, with Bangladesh, Iran, and the Philippines the leading importers.

Africa's Food-Supply Crisis
Some countries that previously depended on imported grain have become self-sufficient in recent years. Higher productivity generated by the green revolution is primarily responsible for reducing dependency on imports, especially in Asia.

In contrast, sub-Saharan Africa is losing the race to keep food production ahead of population growth. By all estimates, the problems will grow worse. Production of most food crops is lower today in Africa than in the 1960s. Agriculture in sub-Saharan Africa can feed little more than half of the region's population.

The problem is particularly severe in the Horn of Africa, and in the Sahel region.

(364)
With rapid population growth, pastoral nomad herd size increased beyond the capacity of the land to support them. Farmers overplanted, exhausting soil nutrients, and reduced fallow time, during which unplanted fields can recover. Soil erosion increased after most of the remaining trees were cut for wood and charcoal, used for urban cooking and heating.

Government policies have aggravated the food-shortage crisis. To make food affordable for urban residents, governments keep agricultural prices low. Constrained by price controls, farmers are unable to sell their commodities at a profit and therefore have little incentive to increase productivity.

Key Terms

Agribusiness (p.338)
Agriculture (p.333)
Cereal grain (p.347)
Chaff (p.345)
Combine (p.350)
Commercial agriculture (p.335)
Crop (p.333)
Crop rotation (p.346)
Desertification (p.359)
Double cropping (p.345)
Grain (p.350)
Green revolution (p.360)
Horticulture (p.354)
Hull (p.345)
Intensive subsistence agriculture (p.345)
Milkshed (p.349)
Paddy (p.345)
Pastoral nomadism (p.343)
Pasture (p.344)
Plantation (p.355)

Prime agricultural land (p.338)
Ranching (p.351)
Reaper (p.350)
Ridge tillage (p.357)
Sawah (p.345)
Seed agriculture (p.334)
Slash-and-burn agriculture (p.339)
Shifting cultivation (p.339)
Spring wheat (p.350)
Subsistence agriculture (p.335)
Sustainable agriculture (p.357)
Swidden (p.341)
Thresh (p.345)
Transhumance (p.344)
Truck farming (p.355)
Vegetative planting (p.334)
Wet rice (p.345)
Winnow (p.345)
Winter wheat (p.350)

Chapter 11. Industry

The recent success of Japan, South Korea, Taiwan, and other Asian countries is a dramatic change from the historic dominance of world industry by Western countries.

Key Issues
1. Where did industry originate?
2. Where is industry distributed?
3. Why do industries have different distributions?
4. Why do industries face problems?

(371)
The title of this chapter, "Industry," refers to the manufacturing of goods in a factory. The word is appropriate, because it also means persistence or diligence in creating value. Industry is much more highly clustered in *space* than is agriculture. Two connections are critical in determining the best location for a factory: *where* the markets for the product are located, and where the resources needed to make the product are located. A generation ago, industry was highly clustered in a handful of more developed countries, but industry has diffused to less developed countries. Geographers identify a community's assets that enable it to compete successfully for industries, as well as handicaps that must be overcome to retain older companies.

Key Issue 1. Where Did Industry Originate?
- **The Industrial Revolution**
- **Diffusion of the Industrial Revolution**

The modern concept of industry—meaning the manufacturing of goods in a factory—began in the United Kingdom in the late 1700s. This process of change is called the **Industrial Revolution**, discussed in Chapter 2 as a cause of population growth between 1750 and 1950. This section examines the sequence of changes that occurred with the Industrial Revolution and its diffusion from the United Kingdom to the rest of the world.

The Industrial Revolution
From its beginnings in the north of the United Kingdom around 1750, the Industrial Revolution diffused to Europe and North America in the nineteenth century and to the rest of the world in the twentieth century. The Industrial Revolution resulted in new social, economic, and political inventions, not just industrial ones. Prior to the Industrial Revolution, industry was geographically dispersed across the landscape. Home-based manufacturing was known as the **cottage industry** system. The Industrial Revolution was the collective invention of hundreds of mechanical devices.

(372)
Diffusion of the Industrial Revolution
The iron industry was first to increase production through extensive use of (James) Watt's steam engine, plus other inventions. The textile industry followed. From these two pioneering industries, new industrial techniques diffused during the nineteenth century.

Diffusion from the Iron Industry
Iron ore is mined from the ground. The ore is not in a useful form for making tools, so it has to be smelted (melted) in a blast furnace (blasted with air to make its fires burn hotly). Henry Cort . . . patented two processes, known as puddling and rolling, in 1783, . . . to remove carbon and other impurities. The combination of Watt's engine and Cort's iron purification process increased iron-manufacturing capability. The needs of the iron industry in turn generated innovations in coal mining, engineering, transportation, and other industries. These inventions in turn permitted the modernization of other industrial activities.

Coal. Wood, the main energy source prior to the Industrial Revolution, became increasingly scarce because it was needed for construction of ships, buildings, and furniture, as well as for heat. High-energy coal . . . was plentiful. Because of the need for large quantities of bulky, heavy coal, the iron industry's geographic pattern changed from dispersed to clustered. These factories clustered at four locations. Each site was near a productive coalfield.

Engineering. In 1795 James Watt decided to go into business for himself rather than serve as a consultant to industrialists. He and Matthew Boulton established the Soho Foundry at Birmingham, England, and produced hundreds of new machines. From this operation came our modern engineering and manufacture of machine parts.

Transportation. The new engineering profession made its biggest impact on transportation, especially canals and railways.

(373)
In 1759 Francis Egerton, the second Duke of Bridgewater, decided to build a canal between Worsley and Manchester. This feat launched a generation of British canal construction. The canals soon were superseded by the invention of another transportation system, the railway, or "iron horse." The railway was not invented by one individual, but through teamwork. Two separate but coordinated engineering improvements were required: the locomotive, and iron rails for it to run on. The first public railway was opened between Stockton and Darlington in the north of England in 1825.

Diffusion from the Textile Industry
A series of inventions between 1760 and 1800 transformed textile production from a dispersed cottage industry to a concentrated factory system. Richard Arkwright . . . improved the process of spinning yarn. He produced a spinning frame in 1768 . . . then . . . a process for carding (untwisting the fibers prior to spinning). These two operations required more power than human beings could provide. The textile industry joined the iron industry early in adopting Watt's steam engine.

From the clothing industry's need for new bleaching techniques emerged another industry that is characteristic of the Industrial Revolution: chemicals.

Chemicals. The traditional method of bleaching cotton involved either exposing the fabric to the Sun or boiling it, . . . first . . . in a solution of ashes and then in sour milk. In 1746 John Roebuck and Samuel Garbett established a factory in which sulfuric acid, obtained from burning coal, was used instead of sour milk. In 1798 Charles Tennant, who produced a bleaching powder made from chlorine gas and lime, a safer product than sulfuric acid. Sulfuric acid was also used to dye clothing. Combined with various metals, sulfuric acid produced another acid, called vitriol, the color of which varied with the metal, . . . blue with copper, green with iron, and white with zinc. Natural-fiber cloth, such as cotton and wool, is now combined with chemically produced synthetic fibers, . . . made from petroleum or coal derivatives. Today the largest textile factories are owned by chemical companies.

(374)
Food Processing. Another industry derived from the chemical industry is food processing. Canning . . . requires high temperature over time . . . some four to five hours, depending on the product. This is where chemical experiments contributed. Calcium chloride was added to the water, raising its boiling temperature from 100°C to 116°C (212°F to 240°F). This reduced the time for proper sterilization to only 25 to 40 minutes. Consequently, production of canned foods increased tenfold.

Diffusion from the United Kingdom
Britain's Crystal Palace became the most visible symbol of the Industrial Revolution, . . . built to house the 1851 . . . "Great Exhibition of the Works of Industry of All Nations." When Queen Victoria opened the Crystal Palace, the United Kingdom was the world's dominant industrial power. From the United Kingdom, the Industrial Revolution diffused eastward through Europe and westward across the Atlantic Ocean to North America. From these places, industrial development continued diffusing to other parts of the world.

Diffusion to Europe. Europeans developed many early inventions of the Industrial Revolution in the late 1700s. The Belgians led the way in new coal-mining techniques, the French had the first coal-fired blast furnace for making iron, and the Germans made the first industrial cotton mill. Political instability delayed the diffusion of the Industrial Revolution in Europe.

Europe's political problems retarded development of modern transportation systems, especially the railway. Railways in some parts of Europe were delayed 50 years after their debut in Britain. The Industrial Revolution reached Italy, the Netherlands, Russia, and Sweden in the late 1800s. Other Southern and Eastern European countries joined the Industrial Revolution during the twentieth century.

Diffusion to the United States. Industry arrived a bit later in the United States than in Western European countries like France and Belgium, but it grew much faster.

(375)
The first U.S. textile mill was built in Pawtucket, Rhode Island, in 1791, by Samuel Slater, a former worker at Arkwright's factory in England. The textile industry grew rapidly after 1808, when the U.S. government imposed an embargo on European trade to avoid entanglement in the Napoleonic Wars. The United States had become a major industrial nation by 1860, second only to the United Kingdom. However, except for textiles, leading U.S. industries did not widely use the new industrial processes . . . until the final third of the nineteenth century. Although industrial development has diffused across Earth's surface, much of the world's industry is concentrated in four regions.

Key Issue 2. Where Is Industry Distributed?
- **North America**
- **Europe**
- **East Asia**

Approximately three-fourths of the world's industrial production is concentrated in four regions: eastern North America, northwestern Europe, Eastern Europe, and East Asia. Agriculture occupies one-fourth of Earth's land area. In contrast, less than 1 percent of Earth's land is devoted to industry.

North America
Manufacturing in North America is concentrated in the northeastern quadrant of the United States and in southeastern Canada. Only 5 percent of the land area of these countries . . . contains one-third of the population and nearly two-thirds of the manufacturing output.

(376)
This manufacturing belt has achieved its dominance through a combination of historical and environmental factors. Early . . . settlement gave eastern cities an advantage . . . to become the country's dominant industrial center.

The Northeast also had essential raw materials . . . and good transportation. The Great Lakes and major rivers . . . were supplemented in the 1800s by canals, railways, and highways.

Industrialized Areas within North America

Within the North American manufacturing belt, several heavily industrialized areas have developed.

New England. The oldest industrial area in the northeastern United States is southern New England.

Middle Atlantic. The Middle Atlantic area . . . has long attracted industries that need proximity to a large number of consumers. Industries that depend on foreign markets or imported raw materials have located near one of this region's main ports.

Mohawk Valley. A linear industrial belt developed in upper New York State along the Hudson River and Erie Canal. Inexpensive, abundant electricity, generated at nearby Niagara Falls, has attracted aluminum, paper, and electrochemical industries to the region.

Pittsburgh–Lake Erie. The area between Pittsburgh and Cleveland is the nation's most important steel-producing area . . . originally concentrated . . . because of its proximity to Appalachian coal and iron ore. When northern Minnesota became the main source of iron ore, the . . . region could bring in ore via the Great Lakes.

Western Great Lakes. The western Great Lakes area extends from Detroit and Toledo . . . to Chicago and Milwaukee, Wisconsin. Chicago . . . is . . . the hub of the nation's transportation network. Road, rail, air, and sea routes converge on Chicago. Automobile manufacturers and other industries . . . locate in the western Great Lakes region to take advantage of this convergence of transportation routes.

St. Lawrence Valley–Ontario Peninsula. Canada's most important industrial area is the St. Lawrence Valley–Ontario Peninsula area. The region has several assets: centrality to the Canadian market, proximity to the Great Lakes, and access to inexpensive hydroelectric power from Niagara Falls.

(377)
Changing Distribution of U.S. Manufacturing
The total number of manufacturing jobs in the United States has changed little during the past three decades. Constancy at the national scale has hidden sharp changes in the distribution within the United States. The Northeast has lost two million manufacturing jobs during the last three decades, whereas the South and West have grown by one million.

Industry has grown in areas outside the main U.S. manufacturing belt, (with) . . . steel, textiles, tobacco products, and furniture industries . . . in the South. Along the Gulf Coast are oil refining, petrochemical manufacturing, food processing, and aerospace product manufacturing.

The Southeast attracts manufacturers that seek a location where few workers have joined labor unions . . . (in) **right-to-work states.**

On the West Coast, Los Angeles first built its manufacturing base on aircraft and military equipment, but it is now a center for low-tech manufacturing: clothing and textile production, . . . furniture, . . . and food processing. San Diego has . . . naval operations. High-tech manufacturing is located . . . in the San Francisco Bay area . . . and Seattle, home of . . . Boeing and . . . Microsoft.

Europe
The Western European industrial region appears as one region on a world map. In reality, four distinct districts have emerged, primarily because European countries competed with one another to develop their own industrial areas. Eastern Europe has six major industrial regions. Four are entirely in Russia, one is in Ukraine, and one is southern Poland and northern Czech Republic.

(378)

Western Europe

Western Europe's four main industrial districts are the Rhine–Ruhr Valley, the mid-Rhine, the United Kingdom, and northern Italy. Each of these areas is divided into subareas.

Rhine–Ruhr Valley. Western Europe's most important industrial area is the Rhine–Ruhr Valley . . . in northwestern Germany . . . Belgium, France, and the Netherlands. Within the region, industry is dispersed rather than concentrated in one or two cities. No individual city has more than one million inhabitants. The Rhine divides into multiple branches as it passes through the Netherlands. The city of Rotterdam is near to where several major branches flow into the North Sea. This location at the mouth of Europe's most important river has made Rotterdam the world's largest port. Iron and steel manufacturing has concentrated in the Rhine–Ruhr Valley because of proximity to large coalfields. Access to iron and steel production stimulated the location of other heavy-metal industries, such as locomotives, machinery, and armaments.

Mid-Rhine. The second most important industrial area in Western Europe includes southwestern Germany, northeastern France, and the small country of Luxembourg. In contrast to the Rhine–Ruhr Valley, the German portion of the Mid-Rhine region lacks abundant raw materials, but it is at the center of Europe's most important consumer market.

The French portion of the Mid-Rhine region—Alsace and Lorraine—contains Europe's largest iron-ore field and is the production center for two-thirds of France's steel. Tiny Luxembourg is also one of the world's leading steel producers, because the Lorraine iron-ore field extends into the southern part of the country.

United Kingdom. The Industrial Revolution originated in the Midlands and northern England and southern Scotland, . . . in part because those areas contained a remarkable concentration of innovative engineers and mechanics during the late eighteenth century.

(379)

The United Kingdom lost its international industrial leadership in the twentieth century. Britain was saddled with . . . outmoded and deteriorating factories and . . . their "misfortune" of winning World War II. The losers, Germany and Japan, . . . received American financial assistance to build modern factories, replacing those destroyed during the war. The United Kingdom expanded industrial production in the late twentieth century by attracting new high-tech industries that serve the European market. Japanese companies have built more factories in the United Kingdom than has any other European country. Today British industries are more likely to locate in southeastern England near . . . the country's largest concentrations of population and wealth and . . . the Channel Tunnel.

Northern Italy. A fourth European industrial region of some importance lies in the Po River Basin of northern Italy. Modern industrial development in the Po Basin began with establishment of textile manufacturing during the nineteenth century . . . because of two key assets: numerous workers . . . and inexpensive hydroelectricity.

Eastern Europe

Four of Eastern Europe's six industrial districts—Central industrial district, St. Petersburg, Eastern Ukraine, and Silesia—became manufacturing centers in the 1800s. The Volga and Urals regions were established by the Communists during the twentieth century. Russia also contains another major industrial region, Kuznetsk, in the eastern or Asian portion of the country.

Central Industrial District. Russia's oldest industrial region is centered around Moscow. Situated near the country's largest market . . . the Central industrial district produces one-fourth of Russian industrial output.

St. Petersburg Industrial District. Railways were built in the St. Petersburg area several decades earlier than in the rest of Russia. St. Petersburg area specializes in shipbuilding and . . . also produces goods that meet the needs of the local market, such as food-processing, textiles, and chemicals.

Eastern Ukraine Industrial District. One of the world's largest coal reserves, . . . large deposits of iron ore, manganese, and natural gas, make this region Eastern Europe's largest producer of pig iron and steel.

The Volga Industrial District. The district grew rapidly during World War II, when . . . occupied by the invading German army. The Volga district contains Russia's largest petroleum and natural gas fields.

(380)

The Urals Industrial District. The Ural mountain range contains more than 1,000 types of minerals, the most varied collection found in any mining region in the world. Although well endowed with metals, industrial development is hindered by a lack of nearby energy sources. Russia controls nearly all the Urals minerals, although the southern portion of the region extends into Kazakhstan.

Kuznetsk Industrial District. Kuznetsk is Russia's most important manufacturing district east of the Ural Mountains. The region contains the country's largest reserves of coal and an abundant supply of iron ore.

Silesia. Outside the former Soviet Union, Eastern Europe's leading manufacturing area is in Silesia, which includes southern Poland and the northern Czech Republic. It is an important steel production center because it is near coalfields, although iron ore must be imported.

East Asia
East Asia is the most heterogeneous industrial region from the perspective of level of development. South Korea and Taiwan . . . have per capita GDPs substantially above that of China but substantially below that of Japan. China has abundant reserves of coal, iron ore, and other important minerals, but Japan and the other East Asian countries have few natural resources.

The region has become a major exporter of consumer goods. East Asia has taken advantage of its most abundant resource: a large labor force. Japan became an industrial power in the 1950s and 1960s, initially by producing goods . . . in large quantity at cut-rate prices. Prices were kept low, despite high shipping costs. Aware that South Korea, Taiwan, and other Asian countries were building industries based on even lower-cost labor, Japan started training workers for highly skilled jobs. As a result, during the 1970s and 1980s . . . the country became the world's leading manufacturer of automobiles, ships, cameras, stereos, and televisions.

In recent years China has become a major manufacturer of steel, farm machinery, and construction materials . . . (and) clothing. A potentially enormous domestic market (and) . . . low-cost labor has attracted foreign investment in China's industries. As in other regions, industry is not distributed uniformly within East Asia.

(381)
Although industry is located elsewhere in the world, the four industrial regions of North America, Western Europe, Eastern Europe, and Japan account for most of the world's industrial production.

Key Issue 3. Why Do Industries Have Different Distributions?
- **Situation factors**
- **Site factors**
- **Obstacles to optimum location**

Industry seeks to maximize profits by minimizing production costs. A company ordinarily faces two geographical costs: situation and site. **Situation factors** involve transporting materials to and from a factory. **Site factors** result from the unique characteristics of a location. Land, labor, and capital . . . vary among locations. The particular combination of critical factors varies among firms.

Situation Factors

A manufacturer tries to locate its factory as close as possible to both buyers and sellers. If the cost of transporting the product exceeds the cost of transporting inputs, then optimal plant location is as close as possible to the customer. Conversely, if inputs are more expensive to transport, a factory should locate near the source of inputs.

Location near Inputs

If the weight and bulk of any one input is particularly great, the firm may locate near the source of that input to minimize transportation cost.

Copper Industry. The North American copper industry is a good example of locating near the source of heavy, bulky inputs to minimize transportation cost. Copper concentration is a **bulk-reducing industry**, an economic activity in which the final product weighs less than its inputs.

Two-thirds of U.S. copper is mined in Arizona, so the state also has most of the concentration mills and smelters. Arizona also contains refineries, but others are located in coastal states. The East Coast refineries import much of their copper. In general, metal processors such as the copper industry also try to locate near economical electrical sources.

(382)

Steel Industry. The U.S. steel industry . . . demonstrates how locations change when the source and cost of raw materials change. The main inputs for steel production are iron ore and coal. The U.S. steel industry concentrated during the mid-nineteenth century around Pittsburgh in southwestern Pennsylvania, where iron ore and coal were both mined. Steel mills were built during the late 1800s around Lake Erie. The locational shift was largely influenced by the discovery of rich iron ore in the Mesabi Range, . . . in northern Minnesota.

New steel mills were located farther west around 1900, near the southern end of Lake Michigan. Changes in steelmaking required more iron ore in proportion to coal. Most large U.S. steel mills built during the first half of the twentieth century were located . . . near the East and West coasts. Iron ore increasingly came from other countries. Further, scrap iron and steel—widely available in the large metropolitan areas of the East and West coasts—had become an important input in the steel-production process.

(383)

Recently, more steel plants have closed than opened in the United States. Among the survivors, plants around southern Lake Michigan and along the East Coast have significantly increased their share of national production. This success derives primarily from market access rather than input access. In contrast with the main historical locational factor—transportation cost of raw materials—successful steel mills today are located increasingly near major markets. The growth of steel minimills also demonstrates the increasing importance of access to markets rather than to inputs.

Location near Markets

The cost of transporting goods to consumers is a critical locational factor for three types of industries: bulk-gaining, single-market, and perishable.

Bulk-gaining Industries. A **bulk-gaining industry** makes something that gains volume or weight during production. Soft-drink bottling is a good

130

example. Major soft-drink companies like Coca-Cola and Pepsico manufacture syrups according to proprietary recipes and ship them to bottlers in hundreds of communities. Because water is available where people live, bottlers can minimize costs by producing soft drinks near their consumers instead of shipping water (their heaviest input) long distances. Most major bottlers of beer, such as Anheuser-Busch and Miller, follow a similar pattern. Scotch whiskey is another weight-gaining product, but its spatial distribution differs from that of soft drinks. It does not have sufficient consumers to justify a bottling plant in each city.

More commonly, bulk-gaining industries manufacture products that gain volume rather than weight.

(384)
Common fabricated products include televisions, refrigerators, and motor vehicles. If the fabricated product occupies a much larger volume than its individual parts, then the cost of shipping the final product to consumers is likely to be a critical factor. As an example, motor vehicles are assembled at about 60 plants in the United States. The distribution of motor-vehicle assembly plants has changed during the past two decades.

(385)
Proximity to markets is still the most critical factor, but the market has changed, and with it the optimal location for factories. The market has changed through greater diversity of products.

(386)
If a company has a product that is made at only one plant, and the critical locational factor is to minimize the cost of distributing it to U.S. and Canadian consumers, then the optimal factory location is in the U.S. interior rather than on the East or West Coast.

Single-Market Manufacturers. Single-market manufacturers make products sold primarily in one location, so they also cluster near their markets. For example, several times a year, buyers . . . come to New York from all over the United States to select high-style apparel. Manufacturers of fashion clothing then receive large orders for certain garments to be delivered in a short time. Consequently, high-style clothing manufacturers concentrate around New York.

Manufacturers in turn demand rapid delivery of specialized components, such as clasps, clips, pins, and zippers. The specialized component manufacturers, therefore, also concentrate in New York. The manufacturers of parts for motor vehicles are also specialized manufacturers. The clustering of parts manufacturers around their customers—the new Japanese-operated U.S. assembly plants—clearly illustrates the adoption of just-in-time.

Perishable Products. To deliver their products to consumers as rapidly as possible, perishable-product industries must be located near their markets. Processors of fresh food into frozen, canned, and preserved products can locate far from their customers. The daily newspaper is an example of a product other than food that is highly perishable because it contains dated information. In European countries, national newspapers are printed in the largest city during the evening and delivered by train throughout the country overnight. Until recently, publishers considered the United States to be too large to make a national newspaper feasible. With satellite technology, however, *The New York Times*, *Wall Street Journal*, and *USA Today* have moved in the direction of national delivery.

(387)
Ship, Rail, Truck, or Air?
Firms seek the lowest-cost mode of transport, but the cheapest of the four alternatives changes with the distance that goods are being sent. The farther something is transported, the lower is the cost per

kilometer (or mile). The cost per kilometer decreases at different rates for each of the four modes, because the loading and unloading expenses differ for each mode. Trucks are most often used for short-distance delivery and trains for longer distances. Ship transport is attractive for very long distances. Air is normally the most expensive alternative for all distances, but an increasing number of firms transport by air to ensure speedy delivery of small-bulk, high-value packages. Late at night, planes filled with packages are flown to a central hub airport . . . then . . . flown to airports nearest their destination, transferred to trucks, and delivered the next morning.

Break-of-Bulk Points. Regardless of transportation mode, cost rises each time that inputs or products are transferred from one mode to another. Many companies that use multiple transport modes locate at a break-of-bulk point. A **break-of-bulk point** is a location where transfer among transportation modes is possible. Important break-of-bulk points include seaports and airports.

Site Factors
Locations near markets or break-of-bulk points have become more important than locations near raw materials for firms in developed countries. However, situation factors do not explain the growing importance of Japanese and other Asian manufacturers. Japan . . . is far from . . . markets. The cost of conducting business varies among locations, depending on the cost of three production factors: land, labor, and capital.

(388)
Land
The land needed to build one-story factories is more likely to be available in suburban or rural locations. Also, land is much cheaper in suburban or rural locations than near the center city. Industries may be attracted to specific parcels of land that are accessible to energy sources.

Prior to the Industrial Revolution, . . . running water and burning of wood were the two most important sources of energy. When coal became . . . dominant . . . in the late eighteenth century, . . . industry began to concentrate in fewer locations. Electricity became an important source of energy for industry in the twentieth century. Although large industrial users usually pay a lower rate than do home consumers . . . industries with a particularly high demand for energy may select a location with lower electrical rates.

The aluminum industry, for example, requires a large amount of electricity. The first aluminum plant was located near Niagara Falls. Aluminum plants have been built near other sources of inexpensive hydroelectric power, including the Tennessee Valley and the Pacific Northwest. Industry may also be attracted to a particular location because of amenities at the site.

Labor
A **labor-intensive industry** is one in which labor cost is a high percentage of expense. Some labor-intensive industries require highly skilled labor to maximize profit, whereas others need less skilled, inexpensive labor.

Textile and Clothing Industries. Textile and clothing production are prominent examples of labor-intensive industries that generally require less skilled, low-cost workers.

(389)

The global distributions of spinning, weaving, and sewing plants are not the same, because the three steps are not equally labor-intensive. (Some) natural fiber . . . manufacturing is concentrated in countries where the principal input—cotton—is grown. The other major natural fiber—wool—is not typically produced in proximity to sheep farms. Synthetic fibers . . . account for an increasing share of textile production. Less developed countries now account for about half of global production of synthetic fibers. Weaving is more likely to locate in LDCs, because labor is a higher percentage of total production cost. Asian countries have become major fabric producers because lower labor cost offsets the expense of shipping inputs and products long distances.

(390)
Most of the world's cotton clothing, such as shirts, trousers, and underwear, still is produced in the MDCs of Europe and North America. Shirt production has declined more than one-third in the United States and Europe during the past quarter century, while remaining about the same in the LDCs.

U.S. Textile and Clothing Industries. U.S. textile weavers and clothing manufacturers have changed locations to be near sources of low-cost employees. U.S. textile and clothing firms were concentrated in the Northeast during the nineteenth century . . . (employing) European immigrants. Workers began to demand better working conditions and higher wages around 1900. Their claim was bolstered by tragic events, such as the 1911 Triangle Shirtwaist Company fire in New York City. Faced with union demands for higher salaries in the Northeast, cotton textile and clothing manufacturers moved to the Southeast, where people were willing to work longer hours for lower wages.

(391)
Although they earned less than their northeastern counterparts, southeastern workers cooperated because wages were higher than those paid for other types of work in the region. Cotton textile and clothing manufacturing in the United States is now located in the Appalachian Mountains and Piedmont of the Southeast.

But the clothing industry has not completely abandoned the Northeast.

(392)
Wool clothing, such as knit outerwear, requires more skill to cut and assemble the material, and skilled textile workers are more plentiful in the Northeast.

Skilled Labor Industries. Companies may become more successful by paying higher wages for skilled labor than by producing an inferior product made by lower-paid, less skilled workers. Computer manufacturers have concentrated in the highest-wage regions in the United States. These regions have a large concentration of skilled workers because of proximity to major university centers. Traditionally in large factories, each worker was assigned one specific task to perform repeatedly. Some geographers call this approach **Fordist**, because the Ford Motor Company was one of the first to organize its production this way. Relatively skilled workers are needed to master the wider variety of assignments given them under more flexible **post-Fordist** work rules.

Capital
The U.S. motor vehicle industry concentrated in Michigan early in the twentieth century largely because this region's financial institutions were more willing than eastern banks to lend money to the industry's pioneers. The most important factor in the clustering of high-tech industries in California's Silicon Valley—even more important than proximity to skilled labor—was the availability of capital.

(393)
One-fourth of all capital in the United States is spent on new industries in the Silicon Valley. Financial institutions in many LDCs are short of funds, so new industries must seek loans from banks

in MDCs. Local and national governments increasingly attempt to influence the location of industry by providing financial incentives. These include grants, low-cost loans, and tax breaks.

Obstacles to Optimum Location

The location that a firm chooses cannot always be explained by situation and site factors. Many industries have become "footloose," meaning they can locate in a wide variety of places. An industry may be especially footloose if it uses facsimile machines, electronic mail, the Internet, and other communications systems to move inputs and products.

Key Issue 4. Why Do Industries Face Problems?
- **Industrial problems from a global perspective**
- **Industrial problems in more developed countries**
- **Industrial problems in less developed countries**

Each government defines problems of industrial development from its own local perspective. But geographers point out that diverse local constraints on industrial growth faced by individual countries are related to conditions in the global economy.

Industrial Problems from a Global Perspective

Global capacity to produce manufactured goods has increased more rapidly than demand for many products.

Stagnant Demand

From the Industrial Revolution's beginnings in the late 1700s until the 1970s, industrial growth in more developed countries was fueled by long-term increases in population and wealth.

(394)
The growth formula was simple: More people with more wealth demanded more industrial goods.

However, demand for many manufactured goods has slowed in MDCs during the past quarter century. More developed countries now have little, if any, population growth. Wages have not risen as fast as prices during the past two decades. Demand has also been flat for many consumer goods in MDCs because of market saturation. Industrial output is also stagnant because of the increasing quality of products. With the improved quality and higher prices, North American consumers are holding on to their vehicles longer than at any time since the 1940s, when production was halted for nearly four years because of World War II. Changing technology has resulted in declining demand for some industrial products. For example, the global steel demand is less than in the mid-1970s. Today's typical motor vehicle uses one-fourth less steel than those built a quarter century ago.

Increased Capacity Worldwide

While demand for products such as steel has stagnated during the past quarter century, global capacity to produce them has increased. Higher industrial capacity is primarily a result of two trends: the global diffusion of the Industrial Revolution and the desire by individual countries to maintain their production despite a global overcapacity. Historically, manufacturing was concentrated in a few locations. Industrial growth through increased international sales was feasible when most of the world was organized into colonies and territories controlled by MDCs. Few colonies remain in the world today, and nearly every independent country wants to establish its own industrial base. The steel industry illustrates the changing distribution.

(395)
The overall level of world steel production in 2000 remained virtually the same as in 1975, but the proportion changed significantly among various regions. Steel production . . . in MDCs has declined in two decades from 90 percent to 65 percent, while the LDCs have increased from 10 percent to 40

percent of the world's output. This global diffusion of steel mills has allowed capacity to exceed demand by a wide margin.

(396)
Steel mills in many countries receive substantial government financial support to remain open. If the mills closed, governments would have to pay unemployment compensation to laid-off workers and deal with the social problems of increased unemployment. Maintaining a steel industry also ensures a domestic steel source in times of crisis.

Industrial Problems in More Developed Countries
Countries at all levels of development face a similar challenge: to make their industries competitive in an increasingly integrated global economy. Each state faces distinctive geographical issues in ensuring that their industries compete effectively.

Impact of Trading Blocs
Industrial competition in the more developed world increasingly occurs not among individual countries, but within regional **trading blocs**. The three most important trading blocs are the Western Hemisphere, Western Europe, and East Asia. Within each bloc, countries cooperate in trade. Each bloc then competes against the other two.

Cooperation within Trading Blocs. The North Atlantic Free Trade Agreement . . . brought Mexico into the free trade zone with the United States and Canada. The three NAFTA partners have been negotiating with other Latin American countries.

The European Union has eliminated most barriers to trade through Western Europe. Cooperation among countries is less formal in East Asia, in part because Japan's neighbors have much lower levels of economic development and unpleasant memories of Japanese military aggression during the 1930s and 1940s. The free movement of most products across the borders has led to closer integration of industries within North America and within Western Europe.

Competition among Trading Blocs. The three trading blocs have promoted internal cooperation, yet they have erected trade barriers to restrict other regions from competing effectively. Faced with a decline in domestic steel production of about one-third during the late 1970s, the U.S. government negotiated a series of voluntary export-restraint agreements with other major steel-producing countries. When these quotas were in effect—from 1982 until 1992—U.S. steel companies spent $24 billion modernizing their plants and buying more efficient equipment.

(397)
The number of steelworkers fell by two-thirds. Steel towns have suffered severely from this decline. Declines in other manufacturing sectors in MDCs have had similar impacts in their communities.

Transnational Corporations. Cooperation and competition within and among trading blocs take place primarily through the actions of large transnational corporations, sometimes called multinational corporations. Initially, transnational corporations were primarily American-owned, but in recent years . . . especially Japan, Germany, France, and the United Kingdom have been active as well. Some transnational corporations locate factories in other countries to expand their markets. Transnational corporations also open factories in countries with lower site factors to reduce their production cost. The site factor that varies most dramatically among countries is labor.

Japanese transnational corporations have been especially active in the United States in recent years. Most plants have been located in a handful of interior states, including Ohio, Indiana, Kentucky, Michigan, Tennessee, and Illinois. German transnationals have clustered in the Carolinas.

Disparities within Trading Blocs

One country or region within a country may have lower levels of income and amenities because it has less industry than other countries or regions within the trading bloc.

Disparities within Western Europe. Europe's most important industrial areas, such as western Germany and northern Italy, are relatively wealthy. Disparities exist at the scale of the individual country as well. Industry is concentrated in the regions most accessible to Western Europe's core of population, wealth, and industry. Germany has had a particularly difficult problem with regional disparities, . . . a legacy of Communist-run East Germany (German Democratic Republic). The European Union, through its European Regional Development Fund, assists its three least industrialized member countries—Greece, Ireland, and Portugal—as well as regions in three other countries that lack industrial investment—Northern Ireland (part of the United Kingdom), southern Italy, and most of Spain. Funds also aid a number of declining industrial areas. A number of Western European countries use incentives to lure industry into poorer regions and discourage growth in the richer regions.

Disparities within the United States. The problem of regional disparity is somewhat different in the United States. The South, historically the poorest U.S. region, has had the most rapid growth since the 1930s, stimulated partly by government policy and partly by changing site factors.

(398)
The Northeast, traditionally the wealthiest and most industrialized region, claims that development in the South has been at the expense of old industrialized communities in New England and the Great Lakes states.

Regional development policies scored some successes as long as national economies were expanding overall, because the lagging regions shared in the national growth. But in an era of limited economic growth for MDCs, governments increasingly questioned policies that strongly encourage industrial location in poorer regions.

Industrial Problems in Less Developed Countries

Knowing that their agriculture-based economies offer limited economic growth, the leaders of virtually every LDC encourage new industry.

Old Problems for LDCs

In some respects, LDCs face obstacles similar to those once experienced by today's MDCs.

Distance from Markets. To minimize geographic isolation, industrializing countries invest scarce resources in constructing and subsidizing transportation facilities.

Inadequate Infrastructure. The LDCs obtain support services by importing advisers and materials from other countries, or by borrowing money to develop domestic sources.

New Problems for LDCs

In addition, industrializing countries now face a new obstacle. Few untapped foreign markets remain to be exploited. New industries must sell primarily to consumers inside their own country. According to principles of economic geography, (in addition to market access), there are two other critical locational factors: access to raw materials and site factors. In fact, new African factories generally are those for which these factors are critical: (1) *raw material access*, (2) *site factors*.

Transnational corporations have been especially aggressive in using low-cost labor in LDCs. Transnational corporations can profitably transfer some work to LDCs, despite greater transportation cost. Operations that require highly skilled workers remain in factories in MDCs. This selective transfer of some jobs to LDCs is known as the **new international division of labor**. Many African countries possess iron ore. But steel, perhaps the most important industry for a less developed country, has had difficulty getting a foothold in Africa. Without cooperation among several small states, steel manufacturing is not likely to develop further in Africa.

Key Terms

Break-of-bulk point (p.387)
Bulk-gaining industry (p.383)
Bulk-reducing industry (p.381)
Cottage industry (p.371)
Fordist (p.392)
Industrial Revolution (p.371)
Labor-intensive industry (p.388)
Maquiladora (p.370)
New international division of labor (p.398)
Post-Fordist (p.392)
Right-to-work state (p.377)
Site factors (p.381)
Situation factors (p.381)
Textile (p.373)
Trading bloc (p.396)

Chapter 12. Services

The regular distribution (of settlements) observed over North America and over other more developed countries is not seen in less developed countries. The regular pattern of settlement in more developed countries reflects where services are provided. In more developed countries the majority of the workers are employed in the tertiary sector of the economy, defined in Chapter 9 as the provision of goods and services to people in exchange for payment. In contrast, less than 10 percent of the labor force in less developed countries provides services.

Key Issues
1. Where did services originate?
2. Why are consumer services distributed in a regular pattern?
3. Why do business services locate in large settlements?
4. Why do services cluster downtown?

(405)

A service is any activity that fulfills a human want or need and returns money to those who provide it. In sorting out where services are distributed in space, geographers see a close link between services and settlements, because services are located in settlements. A settlement is a permanent collection of buildings, where people reside, work, and obtain services. They occupy a very small percentage of Earth's surface, substantially less than 1 percent, but settlements are home to nearly all humans, because few people live in isolation. The optimal location of industry, described in the last chapter, requires balancing a number of site and situation factors; but the optimal location for a service is simply near its customers. On the other hand, locating a service calls for far more precise geographic skills than locating a factory. The optimal location for a service may be a very specific place, such as a street corner. Within more developed countries, larger cities offer a larger scale of services than do small towns, because more customers reside there. As they do for other economic and cultural features, geographers observe trends toward both globalization and local diversity in the distribution of services.

Key Issue 1. Where Did Services Originate?
- **Types of services**
- **Origin of services**
- **Services in rural settlements**

Services are provided in all societies, but in more developed countries a majority of workers are engaged in the provision of services.

Types of Services
The service sector of the economy is subdivided into three types: consumer services, business services, and public services. The first two groups are divided into two subgroups. This division of the service sector has largely replaced earlier approaches that identified tertiary, quaternary, and quinary sectors in various ways.

Consumer Services
Retail services and personal services are the two main types of consumer services.

Retail Services. About one-fifth of all jobs in the United States are in retail services. Within the group, one-fifth of the jobs are in wholesale, one third in . . . food (services).

(406)

Personal Services. Another one-fifth of all jobs in the United States are in personal services. Most of these jobs are in health care or education. The remainder are primarily arts and entertainment and personal care.

Business Services
The principal purpose of business services is to facilitate other businesses. Producer services and transportation . . . are the two main types.

Producer Services. Producer services . . . help people conduct other business. About one-fifth all of U.S. jobs are in producer services.

Transportation and Similar Services. Businesses that diffuse and distribute services are grouped as transportation and information services. In the United States about 7 percent of all jobs are in this group.

Public Services
The purpose of public services is to provide security and protection for citizens and businesses. In the United States about 4 percent of all workers are in public services not already included in other service categories, such as teachers. The distinction among services is not absolute. Individual consumers use business services . . . and businesses use consumer services. Geographers find the classification useful, because the various types of services have different distributions, and different factors influence locational decisions.

Changes in Number of Employees
Between 1970 and 2000 all of the growth in employment in the United States has been in services.

(407)
Producer-service jobs tripled between 1970 and 2000, whereas the number of all service jobs doubled. Professional services . . . quadrupled. Jobs increased by about three-fourths in personal services, retail services, and transportation services, and by about one-third in public services. The personal-service sector has increased rapidly primarily because of a very large increase in the provision of health-care services. Other large increases have been recorded in recreation and entertainment. Services offering personal care have grown more modestly. Within retail services, much of the increase is attributable to the growth of restaurants. Within transportation services, large increases in new technology have been offset by declines in older technology. Public services have expanded much more slowly than other services, despite the widespread misperception that government has expanded massively.

Origin of Services
Services are clustered in settlements. No one knows the precise sequence of events through which settlements were established to provide services. Based on archaeological research, settlements probably originated to provide personal services, especially religion and education, as well as public services such as government and police protection. Transportation, producer, and retail services came later.

Early Personal Services
The early permanent settlements may have been established to offer personal services, specifically places to bury the dead. Having established a permanent resting place for the dead, the group might then install priests at the site to perform the service of saying prayers for the deceased. This would have encouraged the building of structures—places for ceremonies and dwellings. Until the invention of skyscrapers in the late nineteenth century, religious buildings were often the tallest structures in a community. Settlements also may have been places to house families, permitting unburdened males to travel farther and faster in their search for food. Women kept "home and hearth," making household objects, such as pots, tools, and clothing—the origin of industry. The education of children became an important service. . . .

(408)
Making pots and educating children evolved over thousands of years into a wide variety of services . . . which create and store a group's values and heritage and transmit them from one generation to the next. People also needed tools, clothing, shelter, containers, fuel, and other material goods. Men

gathered the materials. Women used these materials to manufacture household objects and maintain their dwellings. The variety of personal services expanded as people began to specialize.

Early Public Services
Public services probably followed the religious activities into early permanent settlements. The group's political leaders also chose to live permanently in the settlement. The settlement likely was a good base from which the group could defend nearby food sources against competitors.

(409)
For defense, the group might surround the settlement with a wall. Thus, settlements became citadels. Although modern settlements no longer have walls, their military and political services continue to be important.

Early Retail and Producer Services
Everyone in settlements needed food, which was supplied by the group through hunting or gathering. Settlements took on a retail-service function. People brought objects and materials they collected or produced into the settlement and exchanged them for items brought by others. The settlement served as neutral ground where several groups could safely come together to trade goods and services. To facilitate this trade, officials in the settlement provided producer services, such as regulating the terms of transactions . . . .

Services in Rural Settlements
(With the development of agriculture, as described in Chapter 10), settlements were surrounded by fields, where people produced most of their food by planting seeds and raising animals rather than by hunting and gathering. Most people in the world still live in rural settlements that have changed little in purpose since ancient times. They are known as clustered rural settlements. Dispersed rural settlements, characteristic of the contemporary North American rural landscape, are characterized by farmers living on individual farms.

Clustered Rural Settlements
A clustered rural settlement typically includes homes, barns, tool sheds, and other farm structures, plus personal services, such as religious structures and schools. In common language such a settlement is called a hamlet or village. The fields must be accessible to the farmers and are thus generally limited to a radius of 1 or 2 kilometers (one-half to 1 mile) from the buildings. In some places, individual farmers own or rent the land; in other places, the land is owned collectively by the settlement or by a lord. Parcels of land . . . may be allocated to specific agricultural activities. Consequently, farmers typically . . . have responsibility for . . . scattered parcels in several fields. This pattern . . . encouraged living in a clustered rural settlement to minimize travel time to the various fields. Traditionally, when the population of a settlement grew too large for the capacity of the surrounding fields, new settlements were established nearby. The establishment of satellite settlements often is reflected in place names.

(410)
Clustered rural settlements are often arranged in one of two types of patterns: circular and linear.

Circular Rural Settlements. The circular form consists of a central open space surrounded by structures. The kraal villages in southern Africa have enclosures for livestock in the center. The German Gewandorf settlement consisted of a core of houses, barns, and churches, encircled by different types of agricultural activities.

Linear Rural Settlements. Linear rural settlements feature buildings clustered along a road, river, or dike to facilitate communications. The fields extend behind the buildings in long, narrow strips. Today, in North America, linear rural settlements exist in areas settled by the French. The French settlement pattern, called long-lot or seigneurial, was commonly used along the St. Lawrence River in Québec and the lower Mississippi River.

Colonial American Clustered Settlements. The first European colonists settled along the East Coast in three regions: New England, the Southeast, and the Middle Atlantic. New England colonists built clustered settlements centered on an open area called a common. Clustered settlements were favored by New England colonists for a number of reasons. Typically, they traveled to the American colonies in a group. The settlement was usually built near the center of the land grant.

(411)
New England settlements were also clustered to reinforce common cultural and religious values. Colonists also favored clustered settlements for defense against Indian attacks. Each villager owned several discontinuous parcels on the periphery of the settlement, to provide the variety of land types needed for different crops. Beyond the fields the town held pastures and woodland for the common use of all residents.

The southeastern colonies were first settled in the 1600s with small, dispersed farms. Then a different style emerged, called a plantation. Plantations grew more profitable in the 1700s when the tobacco and cotton markets expanded and two large sources of labor were identified: indentured whites . . . and black slaves. The plantation's wealthy owner lived in a large mansion. Surrounding the mansion were service buildings.

Dispersed Rural Settlements
Dispersed rural settlements have become more common in the past 200 years, especially in Anglo-America and the United Kingdom, because in more developed societies they are generally considered more efficient than clustered settlements.

Dispersed Rural Settlements in the United States. The Middle Atlantic colonies were settled by a more heterogeneous group of people. Further, most Middle Atlantic colonists came as individuals.

(412)
Dispersed settlement patterns dominated in the American Midwest in part because the early settlers came primarily from the Middle Atlantic colonies. In New England a dispersed distribution began to replace the clustered settlements in the eighteenth century. In part, the cultural bonds that had created clustered rural settlements had weakened. Owning several discontinuous fields had several disadvantages: Eventually people bought, sold, and exchanged land to create large, continuous holdings instead of several isolated pieces. A shortage of land eventually forced immigrants and children to strike out alone and claim farmland on the frontier.

Enclosure Movement. To improve agricultural production, a number of European countries converted their rural landscapes from clustered settlements to dispersed patterns. A prominent example was the enclosure movement in Great Britain, between 1750 and 1850. Because the enclosure movement coincided with the Industrial Revolution, villagers who were displaced from farming moved to urban settlements and became workers in factories and services. The enclosure movement brought greater agricultural efficiency, but it destroyed the self-contained world of village life.

Key Issue 2. Why Are Consumer Services Distributed in a Regular Pattern?
- **Central place theory**
- **Market-area analysis**
- **Hierarchy of services and settlements**

Consumer services and business services do not have the same distributions. Consumer services generally follow a regular pattern based on size of settlements. Business services . . . cluster in specific settlements, creating a specialized pattern.

Central Place Theory

A central place is a market center for the exchange of goods and services by people attracted from the surrounding area. Central places compete against each other. This competition creates a regular pattern of settlements, according to central place theory.

Market Area of a Service

The area surrounding a service from which customers are attracted is the market area or hinterland. To establish the market area, a circle is drawn around the node of service on a map. The closer to the periphery of the circle, the greater is the percentage of consumers who will choose to obtain services from other nodes.

(413)
But circles cause a geometric problem. They either overlap or have gaps between them. Central place theory requires a geometric shape without gaps or overlaps, so circles are out. Squares fit without gaps, but . . . the distance from the center varies among points along a square. To represent a market area, the hexagon is the best compromise between circles and squares.

Size of Market Area

To determine the extent of a market area, geographers need two pieces of information about a service: its range and its threshold.

Range of a Service. The range is the maximum distance people are willing to travel to use a service. The range is the radius of the circle drawn to delineate a service's market area. If firms at other locations compete by providing the service, the range must be modified. The irregularly shaped circle takes in the territory for which the proposed site is closer than competitors.'

The range must be modified further because most people think of distance in terms of time, rather than a linear measure like kilometers or miles. The irregularly shaped circle must be drawn to acknowledge that travel time varies with road conditions.

Threshold of a Service. The second piece of geographic information needed to compute a market area is the threshold, which is the minimum number of people needed to support the service.

(414)
How potential consumers inside the range are counted depends on the product. Developers of shopping malls, department stores, and large supermarkets typically count only higher-income people.

Market-Area Analysis

Retailers and other service providers make use of market-area studies to determine whether locating in the market would be profitable and, if so, the best location within the market area.

Profitability of a Location

The range and threshold together determine whether a good or service can be profitable in a particular location. A store may need a larger threshold and range to attract some of the available customers if competitors are located nearby.

Optimal Location within a Market

According to geographers, the best location is the one that minimizes the distance to the service for the largest number of people.

(415)
Best Location in a Linear Settlement. In a linear community like an Atlantic Ocean resort, the service should be located where half of the customers are to the north and half to the south. It corresponds to the median, which mathematically is the middle point in any series of observations.

What if a different number of customers live in each block of the city?

(416)
To compute the optimal location in these cases, geographers have adapted the gravity model from physics. The gravity model predicts that the optimal location of a service is directly related to the number of people in the area and inversely related to the distance people must travel to access it.

Best Location in a Nonlinear Settlement. Most settlements are more complex than a single main street. Geographers still apply the gravity model to find the best location.

Hierarchy of Services and Settlements
Small settlements are limited to services that have small thresholds, short ranges, and small market areas. Larger settlements provide services having larger thresholds, ranges, and market areas. However, neighborhoods within large settlements also provide services having small thresholds and ranges.

Nesting of Services and Settlements
More developed countries have numerous small settlements with small thresholds and ranges, and far fewer large settlements with large thresholds and ranges. The nesting pattern can be illustrated with overlapping hexagons of different sizes (for) . . . different levels of market area. In his original study, Walter Christaller showed that the distances between settlements in southern Germany followed a regular pattern. He identified seven sizes of settlements (market hamlet, township center, county seat, district city, small state capital, provincial head capital, and regional capital city). Brian Berry has documented a similar hierarchy of settlements in parts of the U.S. Midwest. The principle of nesting market areas also works at the scale of services within cities.

Rank-Size Distribution of Settlements
In many MDCs, geographers observe that ranking settlements from largest to smallest (population) produces a regular pattern or hierarchy. This is the rank-size rule, in which the country's nth-largest settlement is 1/n the population of the largest settlement.

(418)
Several MDCs in Europe follow the rank-size distribution among smaller settlements, but not among the largest ones. The largest settlement in these countries follows the primate city rule. The largest settlement has more than twice as many people as the second-ranking settlement. In France, for example, Paris is a primate city. The primate city in the United Kingdom (is) London.

Many LDCs also follow the primate-city rule. However, in these countries, the rank-size rule tends to fail at other levels in the hierarchy as well. The existence of a rank-size distribution of settlements is not merely a mathematical curiosity. The absence of the rank-size distribution in a less developed country indicates that there is not enough wealth in the society to pay for a full variety of services.

Key Issue 3. Why Do Business Services Locate in Large Settlements?
- **World Cities**
- **Hierarchy of business services**
- **Economic base of settlements**

Business services disproportionately cluster in a handful of settlements, and individual settlements specialize in particular business services.

World Cities
Prior to modern times, virtually all settlements were rural, because the economy was based on the agriculture of the surrounding fields. Providers of personal services and a handful of other types of services met most of the needs of farmers living in the village. Even in ancient times, a handful of

urban settlements provided producer and public services, as well as retail and personal services with large market areas.

Ancient World Cities

Urban settlements . . . may have originated in Mesopotomia . . . and diffused at an early date to Egypt, China, and South Asia's Indus Valley. Or they may have originated independently in each of the four hearths.

Earliest Urban Settlements. Among the oldest well-documented urban settlements is Ur in Mesopotamia (present-day Iraq). Archaeologists have unearthed ruins in Ur that date from approximately 3000 B.C. Ancient Ur was compact, perhaps covering 100 hectares (250 acres), and was surrounded by a wall. The most prominent structure was a temple, known as a ziggurat. Surrounding the ziggurat were residential areas containing a dense network of narrow winding streets and courtyards.

(419)
Recent evidence unearthed at Titris Hoyuk, in present-day Turkey, from about 2500 B.C. suggests that early urban settlements were well-planned communities. Houses varied in size but were of similar design. Houses were apparently occupied by an extended family, because they contained several cooking areas. Titris Hoyuk occupied a 50-hectare (125-acre) site and apparently had a population of about 10,000. The site is especially well-preserved today because after 300 years the settlement was abandoned and never covered by newer buildings.

Ancient Athens. Settlements were first established in the eastern Mediterranean about 2500 B.C., trading centers for the thousands of islands dotting the Aegean Sea and the eastern Mediterranean. They were organized into city-states. The settlement provided the government, military protection, and other public services for the surrounding hinterland. The number of urban settlements grew rapidly during the eighth and seventh centuries B.C. Athens, the largest city-state in ancient Greece, was probably the first city to attain a population of 100,000.

(420)
Ancient Rome. The rise of the Roman Empire encouraged urban settlement. Settlements were established as centers of administrative, military, and other public services, as well as trading and other retail services. The city of Rome—the empire's center for administration, commerce, culture, and all other services—grew to at least a quarter-million inhabitants, although some claim that the population may have reached a million. With the fall of the Roman Empire in the fifth century A.D., urban settlements declined. With the empire fragmented under hundreds of rulers, trade diminished. Large urban settlements shrank or were abandoned. For several hundred years Europe's cultural heritage was preserved largely in monasteries and isolated rural areas.

Medieval World Cities

Urban life began to revive in Europe in the eleventh century as feudal lords established new urban settlements. They gave residents charters of rights to establish independent cities in exchange for fighting for the lord. By the fourteenth century, Europe was covered by a dense network of small market towns serving the needs of particular lords. The largest medieval European urban settlements served as power centers for the lords and church leaders, as well as major market centers. European urban settlements were usually surrounded by walls in medieval times.

(421)
Dense and compact within the walls, medieval urban settlements lacked space for construction, so ordinary shops and houses nestled into the side of the walls and large buildings.

(422)
From the collapse of the Roman Empire until the diffusion of the Industrial Revolution across Europe during the nineteenth century, most of the world's largest cities were in Asia, not Europe. Beijing

144

(China) competed with Constantinople as the world's most populous city for several hundred years, until London claimed the distinction during the early 1800s.

Modern World Cities

In modern times several world cities have emerged where a high percentage of the world's business is transacted and political power is concentrated. These world cities are centers of business services, but they stand at the top of the central place hierarchy in the provision of consumer services, and many also serve as public-service centers. New forms of transportation and communications were expected to reduce the need for clustering of economic activities in large cities. To some extent, economic activities have decentralized, especially manufacturing, but modern inventions reinforce rather than diminish the primacy of world cities in the global economy.

Business Services in World Cities. The clustering of business services in the modern world city is a product of the Industrial Revolution. Factories are operated by large corporations formed to minimize the liability to any individual owner. A board of directors located far from the factory building makes key decisions. Support staff also far from the factory account for the flow of money and materials. This work is done in offices in world cities. World cities offer many financial services to these businesses . . . (and) stock exchanges . . . are located in world cities. Lawyers, accountants, and other professionals cluster in world cities . . . . Advertising agencies, marketing firms, and other services concerned with style and fashion locate in world cities.

Transportation services converge on world cities. They tend to have busy harbors and airports and lie at the junction of rail and highway networks.

Consumer Services in World Cities. Because of their large size, world cities have retail services with extensive market areas, but they may even have more retailers than large size alone would predict. Luxury and highly specialized products are especially likely to be sold there. Personal services of national significance are especially likely to cluster in world cities, in part because they require large thresholds and large ranges, and in part because of the presence of wealthy patrons.

Public Services in World Cities. World cities may be centers of national or international political power. Most are national capitals. Also clustered in the world cities are offices for groups having business with the government. Unlike other world cities, New York is not a national capital. But as the home of the world's major international organization, the United Nations, it attracts thousands of U.N. diplomats and bureaucrats, as well as employees of organizations with business at the United Nations. Brussels is a world city because it is the most important center for European Union activities.

Hierarchy of Business Services

Geographers distinguish four levels of cities that play a major role in the provision of producer and other business services in the global economy . . . a handful of world cities, which can be subdivided into three groups, . . . regional command and control centers, specialized producer-service centers, and dependent centers.

(423)
World Cities. Business services concentrate in disproportionately large numbers in world cities, including law, banking, insurance, accounting, and advertising.

Three world cities stand out in a class of their own: London, New York, and Tokyo. A second tier of major world cities includes Chicago, Los Angeles, and Washington, D.C. in North America, and Brussels, Frankfurt, Paris, and Zurich in Western Europe. Only two of the nine second-tier world cities—São Paulo and Singapore—are in less developed regions.
(424)
A third tier of secondary world cities includes four in North America, seven in Asia, five in Western Europe, four in Latin America, and one each in Africa (Johannesburg) and the South Pacific (Sydney).

Command and Control Centers

The second level of cities—command and control centers—contains the headquarters of many large corporations, . . . concentrations of . . . business services, . . . educational, medical, and public institutions. Two levels of command and control centers can be identified: regional centers and subregional centers. In the United States, examples of regional command centers are Atlanta and Kansas City. Examples of subregional centers are Biloxi and Oklahoma City.

Specialized Producer-Service Centers

The third level of cities, specialized producer-service centers, offers a more narrow and highly specialized variety of services. One group of these cities specializes in the management and R&D activities related to specific industries. A second group . . . specializes as centers of government and education, notably state capitals that also have a major university.

Dependent Centers

The fourth-level cities, dependent centers, provide relatively unskilled jobs and depend for their economic health on decisions made in the world cities, regional command and control centers, and specialized producer-service centers. Four subtypes of dependent centers can be identified in the United States: resort, retirement, and residential centers . . . manufacturing centers . . . industrial and military centers . . . (and) mining and industrial centers.

(425)

Economic Base of Settlements

A settlement's distinctive economic structure derives from its basic industries, which export primarily to consumers outside the settlement. Nonbasic industries are enterprises whose customers live in the same community, essentially consumer services. A community's unique collection of basic industries defines its economic base. A settlement's economic base is important, because exporting by the basic industries brings money into the local economy, thus stimulating the provision of more nonbasic consumer services for the settlement.

A community's basic industries can be identified by computing the percentage of the community's workers employed in different types of businesses. If the percentage is much higher in the local community, (compared to the country), then that type of business is a basic economic activity.

Each type of basic activity has a different spatial distribution. Some settlements have a very high percentage of workers employed in the primary sector, notably mining. The economic base of some settlements is in the secondary sector. Most communities that have an economic base of manufacturing durable goods are clustered between northern Ohio and southeastern Wisconsin, near the southern Great Lakes. Nondurable manufacturing industries, such as textiles, are clustered in the Southeast, especially in the Carolinas.

Specialization of Cities in Different Services

The concept of basic industries originally referred to manufacturing. But in a post-industrial society such as the United States, increasingly the basic economic activities are in business, consumer, or public services. Ó hUallacháin and Reid have documented examples of settlements that specialize in particular types of business services.

(426)

Settlements specializing in public services are dispersed around the country, because these communities typically include a state capital, large university, or military base. Consumer-service settlements include entertainment and recreation centers like Atlantic City, Las Vegas, and Reno, as well as medical centers like Rochester, Minnesota. Business services are concentrated in large metropolitan areas, especially Chicago, Los Angeles, New York, and San Francisco. Although the population of cities in the South and West has grown more rapidly in recent years, Ó hUallacháin and Reid found that cities in the North and East have expanded their provision of business services more rapidly. These cities have moved more aggressively to restructure their economic bases to offset sharp declines in manufacturing jobs.

Distribution of Talent

Individuals possessing special talents are not distributed uniformly among cities. The principal reason enticing talented individuals to cluster in some cities more than others is cultural rather than economic, according to research conducted by Richard Florida. Florida found a significant positive relationship between the distribution of talent and the distribution of diversity in the largest U.S. cities. Attracting talented individuals is important for a city, because these individuals are responsible for promoting economic innovation.

(427)
Key Issue 4. Why Do Services Cluster Downtown?
- **Central business district**
- **Suburbanization of businesses**

Historically, services of all types clustered in the center of the city, commonly called downtown and known to geographers by the more precise term central business district (CBD). Recently services, especially retail, have moved from the CBD to suburban locations.

Central Business District

Often the original site of the settlement, the CBD is compact—less than 1 percent of the urban land area—but contains a large percentage of the shops, offices, and public institutions. The center is the easiest part of the city to reach from the rest of the region and is the focal point of the region's transportation network.

Retail Services in the CBD

Three types of retail services concentrate in the center, because they require accessibility to everyone in the region—shops with a high threshold, shops with a long range, and shops that serve people who work in the center.

Retail Services with a High Threshold. High-threshold shops, such as department stores, traditionally preferred a central location to be accessible to many people. Rents were highest there because this location had the highest accessibility for the most customers. In recent years many high-threshold shops such as large department stores have closed their downtown branches. The customers for downtown department stores now consist of downtown office workers, inner-city residents, and tourists.

Retail Services with a High Range. The second type of shop in the center has a high range. Generally, a high-range shop is very specialized, with customers who patronize it infrequently. Many high-range shops have moved with department stores to suburban shopping malls. These shops can still thrive in some CBDs if they combine retailing with recreational activities.
Entirely new large shopping malls have been built in several downtown areas in North America in recent years. These downtown malls attract suburban shoppers as well as out-of-town tourists because in addition to shops they offer unique recreation and entertainment experiences. A number of cities have preserved their old downtown markets. They may have a high range because they attract customers who willingly travel far to find more exotic or higher-quality products. At the same time, inner-city residents may use these markets for their weekly grocery shopping.

Retail Services Serving Downtown Workers. A third type of retail activity in the center serves the many people who work in the center and shop during lunch or working hours. These businesses sell office supplies, computers, and clothing, or offer shoe repair, rapid photocopying, dry cleaning, and so on. The CBDs in cities outside North America are more likely to contain supermarkets, bakeries, butchers, and other food stores. In contrast to the other two types of retailers, shops that appeal to nearby office workers are expanding in the CBD, in part because the number of downtown office workers has increased and in part because downtown offices require more services.

(428)

Many cities have attempted to revitalize retailing in the CBD and older neighborhoods. One popular method is to ban motor vehicles from busy shopping streets. Shopping streets reserved for pedestrians are widespread in Northern Europe, including the Netherlands, Germany, and Scandinavia.

Producer Services. Offices cluster in the center for accessibility. Despite the diffusion of modern telecommunications, many professionals still exchange information with colleagues primarily through face-to-face contact.

(429)

People in such business services as advertising, banking, finance, journalism, and law particularly depend on proximity to professional colleagues. Services such as temporary secretarial agencies and instant printers locate downtown to be near lawyers, forming a chain of interdependency that continues to draw offices to the center city. A central location also helps businesses that employ workers from a variety of neighborhoods. Firms that need highly specialized employees are more likely to find them in the central area, perhaps currently working for another company downtown.

High Land Costs in the CBD

The center's accessibility produces extreme competition for the limited sites available. As a result, land value in the center is very high. Tokyo's CBD probably contains Earth's most expensive land. Tokyo's high prices result from a severe shortage of buildable land. Buildings in most areas are legally restricted to less than 10 meters in height (normally three stories) for fear of earthquakes. . . . Further, Japanese tax laws favor retention of agricultural land. Tokyo contains 36,000 hectares (90,000 acres) of farmland. Two distinctive characteristics of the central city follow from the high land cost. First, land is used more intensively in the center. Second, some activities are excluded . . . because of the high cost of space.

Intensive Land Use. The intensive demand for space has given the central city a three-dimensional character, pushing it vertically. A vast underground network exists beneath most central cities. The typical "underground city" includes multistory parking garages, loading docks . . . and utility lines. Subways run beneath the streets of larger central cities. Cities such as Minneapolis, Montreal, and Toronto have built extensive pedestrian passages and shops beneath the center. These underground areas segregate pedestrians from motor vehicles and shield them from harsh winter weather.

Skyscrapers. Demand for space in the central city has also made high-rise structures economically feasible. Suburban houses, shopping malls, and factories look much the same from one city to another, but each city has a unique downtown skyline. The first skyscrapers were built in Chicago in the 1880s, made possible by two inventions: the elevator and iron-frame building construction. The first high-rises caused great inconvenience to neighboring structures because they blocked light and air movement. Artificial lighting, ventilation, central heating, and air-conditioning have helped solve these problems. A recent building boom in CBDs of many North American cities is generating problems again; . . . high winds, . . . traffic congestion . . . (and) skyscrapers may prevent sunlight from penetrating to the sidewalks and small parcels of open space. As the Sun and natural air movement are increasingly relied upon again for light and ventilation, the old complaints about high-rises may return. Skyscrapers are an interesting example of "vertical geography." The nature of an activity influences which floor it occupies in a typical high-rise.

(430)

Activities Excluded from the CBD

High rents and land shortage discourage two principal activities in the central area: manufacturing and residence.

Declining Manufacturing in the CBD. The typical modern industry requires a large parcel of land to spread operations among one-story buildings. Suitable land is generally available in suburbs. Port cities in North America and Europe have transformed their waterfronts from industry to commercial and recreational activities. Today's large oceangoing vessels are unable to maneuver in the tight, shallow waters of the old inner-city harbors. Once rotting downtown waterfronts have become major tourist attractions in a number of North American cities, including Boston, Toronto, Baltimore, and San Francisco, as well as in European cities such as Barcelona and London.

Lack of Residents in CBDs. Few people live in U.S. CBDs, because offices and shops can afford to pay higher rents for the scarce space. The shortage of affordable space is especially critical in Europe, because Europeans prefer living near the center city more than Americans do. Abandoned warehouses have been converted into residences in a number of CBDs.

Many people used to live downtown. People have migrated from central areas for a combination of pull and push factors. First, people have been lured to suburbs, which offer larger homes with private yards and modern schools. Second, people have sought to escape from the dirt, crime, congestion, and poverty of the central city.

European CBDs
The central area is less dominated by commercial considerations in Europe than in the United States. In addition to retail and office functions, many European cities display a legacy of low-rise structures and narrow streets, built as long ago as medieval times. Some European cities have tried to preserve their historic core by limiting high-rise buildings and the number of cars. The central area of Warsaw, Poland, represents an extreme example of preservation. The Nazis completely destroyed Warsaw's medieval core during World War II, but Poland rebuilt the area exactly as it had appeared, working from old photographs and drawings. Although constructing large new buildings is difficult, many shops and offices still wish to be in the center of European cities. The alternative to new construction is renovation of older buildings.
(431)
Rents are much higher in the center of European cities than in U.S. cities of comparable size.

Suburbanization of Businesses
Businesses have moved to suburbs: manufacturers . . . because land costs are lower, (and) service providers . . . because most of their customers are there.

Suburbanization of Retailing
Since the end of World War II, downtown sales have not increased, whereas suburban sales have risen at an annual rate of 5 percent. The low density of residential construction discourages people from walking to stores, and restrictive zoning practices often exclude shops from residential areas. Retailing has been increasingly concentrated in planned suburban shopping malls of varying sizes. Corner shops have been replaced by supermarkets in small shopping centers. Malls have become centers for activities in suburban areas that lack other types of community facilities.

Suburbanization of Factories and Offices
Factories and warehouses have migrated to suburbia for more space, cheaper land, and better truck access. Modern factories and warehouses . . . are spread over a single level. Industries increasingly receive inputs and distribute products by truck. Offices that do not require face-to-face contact increasingly are moving to suburbs where rents are much lower than in the CBD.

Key Terms
Basic industries (p.425)

Business services (p.406)

Central business district (CBD)(p.427)

Central place (p.412)

Central place theory (p.412)

City-state (p.419)

Clustered rural settlement (p.409)

Consumer services (p.405)

Chapter 13. Urban Patterns

A large city is stimulating and agitating, entertaining and frightening, welcoming and cold. A city has something for everyone, but a lot of those things are for people who are different from you. Urban geography helps to sort out the complexities of familiar and unfamiliar patterns in urban areas.

Key Issues
1. Where have urban areas grown?
2. Where are people distributed within urban areas?
3. Why do inner cities have distinctive problems?
4. Why do suburbs have distinctive problems?

(439)
Geographers help explain what makes city and countryside different *places*. Urban geographers are interested in the *where* question at two *scales*. First, geographers examine the global distribution of urban settlements. Geographers are also interested in where people and activities are distributed within urban *spaces*. Models have been developed to explain *why* differences occur within urban areas. The major physical, social, and economic contrasts are between inner-city and suburban areas. We all experience the interplay between globalization and local diversity of urban settlements. Many downtowns have a collection of high-rise buildings, towers, and landmarks that are identifiable even to people who have never visited them. On the other hand, . . . suburban houses, streets, schools, and shopping centers look very much alike from one American city to another. In more developed *regions*, people are increasingly likely to live in suburbs. People wish to spread across the landscape to avoid urban problems, but at the same time they want convenient *connections* to the city's jobs, shops, culture, and recreation. Geographers describe where different types of people live and try to explain the reasons for the observed patterns. Although different internal structures characterize urban areas in the United States and elsewhere, the problems arising from current spatial trends are quite similar.

Key Issue 1. Where Have Urban Areas Grown?
* **Urbanization**
* **Defining urban settlements**

As recently as 1800, only 3 percent of Earth's population lived in cities, and only one city in the world—London—had more than one million inhabitants. Two centuries later nearly half of the world's people live in cities, and nearly 400 cities have at least one million inhabitants.

Urbanization
The process by which the population of cities grows, known as **urbanization**, has two dimensions: an increase in the *number* of people living in cities and an increase in the *percentage* of people living in cities.

Increasing Percentage of People in Cities
Within a few years the population of urban settlements will exceed that of rural settlements for the first time in human history. In more developed countries, about three-fourths of the people live in urban areas, compared to about two-fifths in less developed countries. The major exception to the global pattern is Latin America, where the urban percentage is closer to the level of more developed countries. The world map of percentage urban looks very much like the world map of percentage of workers in services. The percentage of urban dwellers is high in more developed countries because over the past 200 years rural residents have migrated from the countryside to work in the factories and services that are concentrated in cities. In more developed countries the process of urbanization that began around 1800 has largely ended, because the percentage living in urban areas simply cannot increase much more. As in more developed countries, people in less developed countries are pushed off the farms by declining opportunities. However, urban jobs are by no means assured in LDCs experiencing rapid overall population growth.

Increasing Number of People in Cities
More developed countries have a higher percentage of urban residents, but less developed countries have more of the large urban settlements (Figure 13–2).

(440)
Six of the 10 most populous cities are currently in LDCs. Urban growth historically has resulted from diffusion of the Industrial Revolution. In 1800 only three of the world's 10 most populous cities were in Europe, . . . and the remainder were in Asia. But in 1900, 9 of the world's 10 most populous cities were in countries that had rapidly industrialized. Tokyo was the only top 10 city then in a preindustrial country. As recently as 1950, 7 of the 10 largest cities in the world remained clustered in MDCs that had industrialized. The rapid growth of cities in the LDCs is a reversal of the historical trend . . . and is not a measure of an improved level of development. Migration from the countryside is fueling 40 percent of the increase. However, half of the growth in the population of urban areas in less developed countries results from high natural increase rates. In Africa, the natural increase rate accounts for three-fourths of urban growth.

Defining Urban Settlements
Defining where urban areas end and rural areas begin is difficult. Geographers and other social scientists have formulated definitions that distinguish between urban and rural areas according to social and physical factors.

(441)
Social Differences between Urban and Rural Settlements
Louis Wirth argued during the 1930s that an urban dweller follows a different way of life from a rural dweller, . . . (and) defined a city as a permanent settlement that has three characteristics: large size, high population density, and socially heterogeneous people.

Large Size. If you live in a rural settlement, you know most of the other inhabitants and may even be related to many of them. In contrast, if you live in an urban settlement, you can know only a small percentage of the other residents. You meet most of them in specific roles. Most of these relationships are contractual.

High Density. According to Wirth, high density also produces social consequences for urban residents. Each person in an urban settlement plays a special role or performs a specific task to allow the complex urban system to function smoothly. At the same time, high density also encourages people to compete for survival in limited space. Social groups compete to occupy the same territory, and the stronger group dominates.

Social Heterogeneity. A person has greater freedom in an urban settlement than in a rural settlement to pursue an unusual profession, sexual orientation, or cultural interest. Regardless of values and preferences, in a large urban settlement individuals can find people with similar interests. Yet despite the freedom and independence of an urban settlement, people may also feel lonely and isolated. Wirth's three-part distinction between urban and rural settlements may still apply in LDCs. But in more developed societies, social distinctions between urban and rural residents have blurred.

Physical Definitions of Urban Settlements
The removal of walls and the rapid territorial expansion of cities have blurred the traditional physical differences. Urban settlements today can be physically defined in three ways: by legal boundary, as continuously built-up area, and as a functional area.

Legal Definition of a City. The term *city* defines an urban settlement that has been legally incorporated into an independent, self-governing unit. In the United States, a city that is surrounded by suburbs is sometimes called a *central city*.

Urbanized Area. An urbanized area consists of a central city plus its contiguous built-up suburbs where population density exceeds 1,000 persons per square mile (400 persons per square kilometer). Approximately 60 percent of the U.S. population lives in urban areas, divided about equally between central cities and surrounding jurisdictions. Working with urbanized areas is difficult because few statistics are available about them. Urbanized areas do not correspond to government boundaries.

Metropolitan Statistical Area. The urbanized area also has limited applicability because it does not accurately reflect the full influence that an urban settlement has in contemporary society. The area of influence of a city extends beyond legal boundaries and adjacent built-up jurisdictions.

(442)
Therefore, we need another definition of urban settlement to account for its more extensive zone of influence. The U.S. Bureau of the Census has created a method of measuring the functional area of a city, known as the **metropolitan statistical area (MSA)**. An MSA includes the following:
1. A central city with a population of at least 50,000
2. The county within which the city is located
3. Adjacent counties with a high population density and a large percentage of residents working in the central city

(443)
The MSAs are widely used because many statistics are published for counties, the basic MSA building block. One problem is that some MSAs include extensive land area that is not urban. The MSAs comprise some 20 percent of total U.S. land area, compared to only 2 percent for urbanized areas. The urbanized area typically occupies only 10 percent of an MSA land area but contains over 90 percent of its population.

The census has also designated smaller urban areas as **micropolitan statistical areas**. These include an urbanized area of between 10,000 and 50,000 inhabitants, the county in which it is found, and adjacent counties tied to the city. About 10 percent of Americans live in a micropolitan statistical area.

Overlapping Metropolitan Areas. Some adjacent MSAs overlap. A county between two central cities may send a large number of commuters to jobs in each.

(444)
In the northeastern United States, large metropolitan areas . . . form one continuous urban complex, extending from north of Boston to south of Washington, D.C. A geographer, Jean Gottmann, named this region Megalopolis, a Greek word meaning great city; others have called it the Boswash corridor. Other continuous urban complexes exist in the United States:
(445)
the southern Great Lakes . . . and southern California. . . . Important examples in other MDCs are the German Ruhr, Randstad in the Netherlands, and Japan's Tokaido.

Within Megalopolis, the downtown areas of individual cities . . . retain distinctive identities. But at the periphery of the urban areas, the boundaries overlap. Once considered two separate areas, Washington and Baltimore were combined into a single metropolitan statistical area after the 1990 census. However, that combination did not do justice to the distinctive character of the two cities, so the census again divided them into two separate MSAs after the 2000 census. At the same time, the Census Bureau has divided other MSAs into two or more metropolitan divisions. For example, Dallas and Fort Worth—long combined into one MSA—are now split into two metropolitan divisions.

Key Issue 2. Where Are People Distributed Within Urban Areas?
- **Three models of urban structure**
- **Use of the models outside North America**

People are not distributed randomly within an urban area. Geographers describe where people with particular characteristics are likely to live within an urban area, and they offer explanations for why these patterns occur.

Three Models of Urban Structure

Sociologists, economists, and geographers have developed three models to help explain where different types of people tend to live in an urban area: the concentric zone, sector, and multiple nuclei models. The three models describing the internal social structure of cities were all developed in Chicago, a city on a prairie. Except for Lake Michigan to the east, few physical features have interrupted the region's growth. The three models were later applied to cities elsewhere in the United States and in other countries.

Concentric Zone Model

The concentric zone model was the first to explain the distribution of different social groups within urban areas. It was created in 1923 by sociologist E. W. Burgess. According to the **concentric zone model**, a city grows outward from a central area in a series of concentric rings.

(446)

The innermost of the five zones is the CBD, where nonresidential activities are concentrated. The CBD is surrounded by the second ring, the zone in transition, which contains industry and poorer-quality housing. Immigrants to the city first live in this zone.

The third ring . . . contains modest older houses occupied by stable, working-class families. The fourth zone has newer and more spacious houses for middle-class families. Finally, Burgess identified a commuters' zone, beyond the continuous built-up area of the city.

Sector Model

A second theory of urban structure, the **sector model**, was developed in 1939 by land economist Homer Hoyt. According to Hoyt, the city develops in a series of sectors, not rings. Certain areas of the city are more attractive for various activities, originally because of an environmental factor or even by mere chance. As a city grows, activities expand outward in a wedge, or sector, from the center. The best housing is therefore found in a corridor extending from downtown to the outer edge of the city. Industrial and retailing activities develop in other sectors, usually along good transportation lines. To some extent the sector model is a refinement of the concentric zone model rather than a radical restatement. Hoyt and Burgess both claimed that social patterns in Chicago supported their model.

Multiple Nuclei Model

Geographers C. D. Harris and E. L. Ullman developed the multiple nuclei model in 1945. According to the **multiple nuclei model**, a city is a complex structure that includes more than one center around which activities revolve. Examples of these nodes include a port, neighborhood business center, university, airport, and park. The multiple nuclei theory states that some activities are attracted to particular nodes while others try to avoid them. For example, a university node may attract well-educated residents, pizzerias, and bookstores, whereas an airport may attract hotels and warehouses.

Geographic Applications of the Models

The three models help us understand where people with different social characteristics tend to live within an urban area. Effective use of the models depends on the availability of data at the scale of individual neighborhoods. Urban areas in the United States are divided into **census tracts**, which contain approximately 5,000 residents and correspond where possible to neighborhood boundaries.

(447)

Every decade, the U.S. Bureau of the Census publishes data summarizing the characteristics of the residents living in each tract.

Social Area Analysis. The spatial distribution of any of these social characteristics can be plotted on a map of the community's census tracts. Social scientists can compare the distributions of characteristics and create an overall picture of where various types of people tend to live. This kind of study is known as *social area analysis.*

Critics point out that the models are too simple and fail to consider the variety of reasons that lead people to select particular residential locations. Critics also question their relevance to contemporary urban patterns . . . because the three models are all based on conditions that existed . . . between the two world wars. But if the models are combined rather than considered independently, they do help geographers explain where different types of people live in a city. The models say that most people prefer to live near others having similar characteristics.

(448)
Putting the three models together, we can identify, for example, the neighborhood in which a childless, high-income, Asian-American family is most likely to live.

Use of the Models outside North America
American urban areas differ from those elsewhere in the world. Social groups in other countries may not have the same reasons for selecting particular neighborhoods.

European Cities
As in the United States, wealthier people in European cities cluster along a sector extending out from the CBD. In Paris, for example, the rich moved to the southwestern hills to be near the royal palace. The preference . . . was reinforced in the nineteenth century during the Industrial Revolution. Factories were built to the south, east, and north along . . . river valleys, but relatively few were built on the southwestern hills. Similar high-class sectors developed in other European cities, typically on higher elevation and near royal palaces. However, in contrast to most U.S. cities, wealthy Europeans still live in the inner rings of the high-class sector, not just in the suburbs. A central location provides proximity to the region's best shops, restaurants, cafes, and cultural facilities. By living in high-density, centrally located townhouses and apartments, wealthy people in Europe do not have large private yards and must go to public parks for open space. To meet the desire for large tracts of privately owned land, some wealthy Europeans purchase abandoned farm buildings in clustered rural settlements for use as second homes on weekends and holidays.

(449)
In the past, poorer people also lived in the center of European cities. Social segregation was vertical: Richer people lived on the first or second floors, while poorer people occupied the dark, dank basements, or they climbed many flights of stairs to reach the attics. During the Industrial Revolution, housing for poorer people was constructed in sectors near the factories.

Today, poorer people are less likely to live in European inner-city neighborhoods. Poor-quality housing has been renovated for wealthy people, or demolished. Building and zoning codes prohibit anyone from living in basements, and upper floors are attractive to wealthy individuals once elevators are installed. Poorer people have been relegated to the outskirts of European cities. Vast suburbs containing dozens of high-rise apartment buildings house the poorer people displaced from the inner city. Many residents of these dreary suburbs are persons of color or recent immigrants from Africa or Asia who face discrimination and prejudice by "native" Europeans. European officials encouraged the construction of high-density suburbs to help preserve the countryside from development and to avoid the inefficient sprawl that characterizes American suburbs. But these policies have resulted in the clustering of people with social and economic problems in remote suburbs rarely seen by wealthier individuals.

Less Developed Countries
In LDCs, as in Europe, the poor are accommodated in the suburbs, whereas the rich live near the center of cities, as well as in a sector extending from the center. The similarity between European and LDC cities is not a coincidence. Most cities in less developed countries have passed through

three stages of development—before European colonization, during the European Colonial period, and since independence.

Pre-Colonial Cities. Before the Europeans established colonies, few cities existed in Africa, Asia, and Latin America, and most people lived in rural settlements.

(450)
Cities were also built in South and East Asia, especially India, China, and Japan. Cities were often laid out surrounding a religious core, such as a mosque in Muslim regions. Government buildings and the homes of wealthy families surrounded the mosque and bazaar. Families with less wealth and lower status located farther from the core, and recent migrants to the city lived on the edge. Commercial activities were arranged in a concentric and hierarchical pattern: Higher-status businesses directly related to religious practices . . . were located closest to the mosque. In the next ring, were secular businesses. Food products were sold in the next ring, then came blacksmiths, basket makers, and potters. A quarter would be reserved for Jews, a second for Christians, and a third for foreigners. When the Aztecs founded Mexico City—which they called Tenochtitlán—the settlement consisted of a small temple and a few huts of thatch and mud . . . west of present-day downtown Mexico City on a hill known as Chapultepec. Forced by other people to leave the hill, they migrated a few kilometers south. Then in 1325 (they moved) to a marshy . . . island in Lake Texcoco. Over the next two centuries the Aztecs conquered the neighboring (territories).

(451)
The Aztecs built elaborate stone houses and temples in Tenochtitlán. The node of religious life was the Great Temple. The main market center, Tlatelolco, was located at the north end of the island. Most . . . merchandise . . . crossed from the mainland to the island by . . . boat. The island itself was laced with canals to facilitate pickup and delivery of people and goods. An aqueduct brought fresh water from Chapultepec.

Colonial Cities. When Europeans gained control of Africa, Asia, and Latin America, they expanded existing cities to provide Colonial services, . . . as well as housing for Europeans who settled in the colony. Fès (Fez), Morocco, consists of two separate and distinct towns—one that existed before the French gained control and one built by the French colonialists.

(452)
On the other hand, the French Colonial city of Saigon, Vietnam (now Ho Minh City), was built by completely demolishing the existing city without leaving a trace. Compared to the existing cities, the European districts typically contain wider streets and public squares, larger houses surrounded by gardens, and much lower density.

Colonial cities followed standardized plans. All Spanish cities in Latin America, for example, were built according to the Laws of the Indies, drafted in 1573. Cities were to be constructed (on) a gridiron street plan centered on a church and central plaza, . . . and neighborhoods centered around smaller plazas with parish churches or monasteries. After the Spanish conquered Tenochtitlán . . . they destroyed the city, and dispersed or killed most of the inhabitants. The city renamed Mexico City, was rebuilt around a main square, called the Zócalo, in the center of the island, on the site of the Aztecs' sacred precinct. The Spanish reconstructed the streets in a grid pattern extending from the Zócalo.

Cities since Independence. Following independence, cities have become the focal points of change in less developed countries. Millions of people have migrated to the cities in search of work. Geographers Ernest Griffin and Larry Ford show that in Latin American cities, wealthy people push out from the center in a well-defined elite residential sector, . . . on either side of a narrow spine that contains offices, shops, and amenities, . . . and . . . services like water and electricity. For example, in Brazil, Rio de Janeiro's high-income people are clustered in the center of the city and to the south, whereas low-income residents are in the northern suburbs. High-income groups are clustered near the center in part because of greater access to services. Physical geography also influences the distribution of social classes within Rio. Residents were

156

attracted to the neighborhoods immediately south of the central area . . . to enjoy spectacular views of the Atlantic Ocean and access to beaches. Low-income households have clustered along the northern edge of the city, where steep mountains have restricted construction of other types of buildings. In Mexico City, Emperor Maximilian (1864–1867) designed a 14-lane, tree-lined boulevard patterned after the Champs-Elysées in Paris.

(453)
The boulevard (now known as the Paseo de la Reforma) extended 3 kilometers southwest from the center to Chapultepec. The Reforma between downtown and Chapultepec became the spine of an elite sector. Physical factors influenced the movement of wealthy people toward the west along the Reforma. Because elevation was higher than elsewhere in the city, sewage flowed eastward and northward away from Chapultepec. Most of Lake Texcoco was drained by a gigantic canal and tunnel project in 1903. However, the lakebed was a less desirable residential location than the west side, because prevailing winds from the northeast stirred up dust storms from the dried-up lakebed. As Mexico City's population grew rapidly during the twentieth century, the social patterns inherited from the nineteenth century were reinforced.

Squatter Settlements. The LDCs are unable to house the rapidly growing number of poor. A large percentage of poor immigrants to urban areas in LDCs live in squatter settlements.

(454)
Squatter settlements have few services, because neither the city nor the residents can afford them. Electricity service may be stolen by running a wire from the nearest power line. In the absence of bus service or available private cars, a resident may have to walk two hours to reach a place of employment.

At first, squatters do little more than camp on the land or sleep in the street. Families then erect primitive shelters with scavenged (materials).

The percentage of people living in squatter settlements, slums, and other illegal housing ranges from 33 percent in São Paulo, Brazil, to 85 percent in Addis Ababa, Ethiopia, according to a U.N. study.

Key Issue 3. Why Do Inner Cities Have Distinctive Problems?
- **Inner-city physical problems**
- **Inner-city social problems**
- **Inner-city economic problems**

Most of the land in urban areas is devoted to residences. Inner cities in the United States contain concentrations of low-income people with a variety of physical, social, and economic problems very different from those faced by suburban residents.

Inner-City Physical Problems
The major physical problem faced by inner-city neighborhoods is the poor condition of the housing, most of which was built before 1940.

Process of Deterioration
As the number of low-income residents increase in the city, the territory they occupy expands. Middle-class families move out of a neighborhood to newer housing farther from the center and sell or rent their houses to lower-income families.

(455)
Filtering. Large houses built by wealthy families in the nineteenth century are subdivided by absentee landlords into smaller dwellings for low-income families. This process of subdivision of houses and occupancy by successive waves of lower-income people is known as **filtering**. Landlords stop maintaining houses when the rent they collect becomes less than the maintenance cost. The building soon deteriorates and grows unfit for occupancy. At this point in the filtering process the owner may abandon the property, because the rents that can be collected are less than

the cost of taxes and upkeep. Governments that aggressively go after landlords to repair deteriorated properties may in fact hasten abandonment, because landlords will not spend money on repairs that they are unable to recoup in rents. These inner-city neighborhoods that housed perhaps 100,000 a century ago contain less than 10,000 inhabitants today. Schools and shops close because they are no longer needed . . . with rapidly declining populations. Through the filtering process, many poor families have moved to less deteriorated houses farther from the center.

Redlining. Some banks engage in **redlining**—drawing lines on a map to identify areas in which they will refuse to loan money. Although redlining is illegal, enforcement of laws against it is frequently difficult. The Community Reinvestment Act requires banks to . . . demonstrate that inner-city neighborhoods within its service area receive a fair share of its loans.

Urban Renewal
North American and European cities have demolished much of their substandard inner-city housing through **urban renewal** programs. The land is then turned over to private developers or to public agencies, . . . to construct new buildings or services.

Public Housing. In the United States, **public housing** is reserved for low-income households, who must pay 30 percent of their income for rent. Public housing accounts for only 2 percent of all dwellings, although it may account for a high percentage of housing in inner-city neighborhoods. In the United Kingdom more than one-third of all housing is publicly owned. Private landlords control only a small percentage of housing in the United Kingdom. Elsewhere in Western Europe, governments typically do not own the housing. Instead, they subsidize construction cost and rent for a large percentage of the privately built housing.

(456)
The U.S. government has also provided subsidies to private developers, but on a much smaller scale than in Europe. Most of the high-rise public-housing projects built in the United States and Europe during the 1950s and early 1960s are now considered unsatisfactory environments for families with children. Some observers claim that the high-rise buildings caused the problem, because too many low-income families are concentrated into a high-density environment. Public-housing authorities have demolished high-rise public-housing projects in recent years in . . . U.S. and European cities.

Cities have also experimented with "scattered-site" public housing, in which dwellings are dispersed throughout the city rather than clustered in a large project. In recent years the U.S. government has stopped funding new public housing. The supply of public housing and other government-subsidized housing diminished by approximately 1 million units between 1980 and 2000. But during the same period, the number of households needing low-rent dwellings increased by more than 2 million. In Britain the supply of public housing, . . . has also declined. The government has forced local authorities to sell some of the dwellings to the residents. But at the same time, the British have expanded subsidies to nonprofit housing associations. Urban renewal has been criticized for destroying the social cohesion of older neighborhoods and reducing the supply of low-cost housing. Most North American and European cities have turned away from urban renewal since the 1970s.

Renovated Housing. In some cases, nonprofit organizations renovate housing and sell or rent them to low-income people. But more often, the renovated housing attracts middle-class people. Most cities have at least one substantially renovated inner-city neighborhood where middle-class people live. In a few cases, inner-city neighborhoods never deteriorated, because the community's social elite maintained them as enclaves of expensive property. The process by which middle-class people move into deteriorated inner-city neighborhoods and renovate the housing is known as **gentrification**. Gentrified inner-city neighborhoods also attract middle-class individuals who work downtown. Cities encourage the process by providing low-cost loans and tax breaks. Public expenditures for renovation have been criticized as subsidies for the middle class at the expense of poor people, who are forced to move . . . because the rents . . . are suddenly too high for them. Cities try to reduce the hardship on poor families forced to move. First, U.S. law requires that they be reimbursed both for moving expenses and for rent increases over a four-year period. Western European countries have similar laws. Second, cities renovate old houses specifically for lower-income families.

Inner-City Social Problems

Beyond the pockets of gentrified neighborhoods, inner cities contain primarily low-income people who face a variety of social problems. Inner-city residents constitute a permanent underclass who live in a culture of poverty.

Underclass

Inner-city residents frequently are referred to as a permanent **underclass** because they are trapped in an unending cycle of economic and social problems.

(457)

Lack of Job Skills. The future is especially bleak for the underclass because they are increasingly unable to compete for jobs. The gap between skills demanded by employers and the training possessed by inner-city residents is widening. Inner-city residents do not even have access to the remaining low-skilled jobs, such as custodians and fast-food servers, because they are increasingly in the distant suburbs.

Homeless. Some of the underclass are homeless. Accurate counts are impossible to obtain, but an estimated one to two million Americans sleep in doorways, on heated street grates, and in bus and subway stations. Homelessness is an even more serious problem in less developed countries. Most people are homeless because they cannot afford housing and have no regular income. Roughly one-third of U.S. homeless are individuals who are unable to cope in society after being released from hospitals or other institutions.

Culture of Poverty

Inner-city residents are trapped as permanent underclass because they live in a culture of poverty. Unwed mothers give birth to two-thirds of the babies in U.S. inner-city neighborhoods, and 90 percent of children in the inner city live with only one parent. Because of inadequate child-care services, single mothers may be forced to choose between working to generate income and staying at home to take care of the children. In principle, government officials would like to see more fathers living with their wives and children, but they provide little incentive for them to do so. If the husband moves back home, his wife may lose welfare benefits, leaving the couple financially worse off together than apart.

Crime. Although drug use is a problem in both the suburbs and rural areas, rates of use in recent years have increased most rapidly in the inner cities. Some drug users obtain money through criminal activities. Violence erupts when two gangs fight over the boundaries between their drug distribution areas. The higher incidence of arrests in low-income African-American areas does not necessarily mean that drug usage is higher or that African-Americans are more involved in drug trafficking than whites. Drug sales in low-density automobile-oriented suburbs may occur discreetly behind closed doors, and arrests may require elaborate undercover operations.

Ethnic and Racial Segregation. Many neighborhoods in the United States are segregated by ethnicity. Even small cities display strong social distinctions among neighborhoods. A family seeking a new residence usually considers only a handful of districts, where the residents' social and financial characteristics match their own.

(458)

Segregation by ethnicity explains voting patterns in many American urban areas.

Inner-City Economic Problems

The concentration of low-income residents in inner-city neighborhoods . . . require public services, but they can pay very little of the taxes to support the services. A city has two choices to close the gap between the cost of services and the funding available from taxes. One alternative is to reduce services. Aside from the hardship imposed on individuals laid off from work, cutbacks in public services also encourage middle-class residents and industries to move from the city.

The other alternative is to raise tax revenues. Because higher tax rates can drive out industries and wealthier people, cities prefer instead to expand their tax base, especially through construction of new CBD projects. Inner-city fiscal problems were alleviated by increasing contributions from the federal government during the 1950s and 1960s. Federal aid to U.S. cities declined by two-thirds during the 1980s when adjusted for inflation. To offset a portion of these lost federal funds, some state governments increased financial assistance to cities.

Annexation
For many cities, economic problems are exacerbated by their inability to annex peripheral land. **Annexation** is the process of legally adding land area to a city.

(459)
Normally, land can be annexed into a city only if a majority of residents in the affected area vote in favor of doing so. Peripheral residents generally desired annexation in the nineteenth century, because the city offered better services. Today, however, cities are less likely to annex peripheral land because the residents prefer to organize their own services rather than pay city taxes for them. As a result, today's cities are surrounded by a collection of suburban jurisdictions. Some of these peripheral jurisdictions were small, isolated towns. Others are newly created communities whose residents wish to live close to the large city but not be legally part of it.

Key Issue 4. Why Do Suburbs Have Distinctive Problems?
* **The peripheral model**
* **Contribution of transportation to suburbanization**
* **Local government fragmentation**

Since 1950, overall population has declined . . . in . . . central cities. The suburban population has grown much faster than the overall population in the United States. In 2000, about 50 percent of Americans lived in suburbs, compared to only 30 percent in central cities and 20 percent in small towns and rural areas. In most (public opinion) polls, more than 90 percent of respondents prefer the suburbs to the inner city. Families with children are especially attracted to suburbs. As incomes rose in the twentieth century, first in the United States and more recently in Western Europe, more families were able to afford to buy suburban homes.

The Peripheral Model
North American urban areas follow what Chauncey Harris (creator of the multiple nuclei model) calls the peripheral model. According to the **peripheral model**, an urban area consists of an inner city surrounded by large suburban residential and business areas tied together by a beltway or ring road.

(460)
The peripheral model points to problems of sprawl and segregation that characterize many suburbs. Around the beltway are nodes of consumer and business services, called **edge cities**. Edge cities originated as suburban residences . . . then shopping malls were built. Now edge cities contain manufacturing centers. Specialized nodes emerge in the edge cities: a collection of hotels and warehouses around an airport, a large theme park, a distribution center near the junction of the beltway and a major long-distance interstate highway.

Density Gradient
As you travel outward from the center of a city, you can watch the decline in the density at which people live. This density change in an urban area is called the **density gradient**. According to the density gradient, the number of houses per unit of land diminishes as distance from the center city increases.

Changes in Density Gradient. Two changes have affected the density gradient in recent years. First, the number of people living in the center has decreased. The density gradient thus has a gap in the center, where few live. Second is the trend toward less density difference within urban areas. The number of people living on a hectare of land has decreased in the central residential areas. At the same time, density has increased on the periphery through construction of apartment and row house. In European cities, density gradient has also been affected by low-income high-rise apartments in the suburbs and by stricter control over construction of detached houses on large lots.

Cost of Suburban Sprawl
U.S. suburbs are characterized by **sprawl**, which is the progressive spread of development over the landscape.

Suburban Development Process. The current system for developing land on urban fringes is inefficient, especially in the United States.

(461)
Developers frequently reject land adjacent to built-up areas in favor of detached isolated sites, depending on the price and physical attributes of the alternatives. The periphery of U.S. cities therefore looks like Swiss cheese, with pockets of development and gaps of open space. Roads and utilities must be extended to connect isolated new developments to nearby built-up areas.

Sprawl also wastes land. Some prime agricultural land may be lost through construction of isolated housing developments; in the interim, other sites lie fallow, while speculators await the most profitable time to build homes on them. The low-density suburb also wastes more energy, especially because the automobile is required for most trips. The supply of land for construction of new housing is more severely restricted in European urban areas . . . by designating areas of mandatory open space. London, Birmingham, and several other British cities are surrounded by **greenbelts**, or rings of open space. New housing is built either in older suburbs inside the greenbelts or in planned extensions to small towns and new towns beyond the greenbelts. Restriction of the supply of land . . . has driven up house prices in Europe.

Suburban Segregation
The modern residential suburb is segregated in two ways. First, residents are separated from commercial and manufacturing activities. Second, . . . a given suburban community is usually built for people of a single social class, with others excluded by virtue of the cost, size, or location of the housing.

(462)
The homogeneous suburb is a twentieth-century phenomenon. In older cities, activities and classes were more likely to be separated vertically rather than horizontally. Poorer people lived on the higher levels or in the basement, the least attractive parts of the building. Once cities spread out over much larger areas, the old pattern of vertical separation was replaced by territorial segregation. Large sections of the city were developed . . . appealing to people with similar incomes and lifestyles.

Zoning ordinances, developed in Europe and North America in the early decades of the twentieth century, encouraged spatial separation. They prevented mixing of land uses within the same district. The strongest criticism of U.S. residential suburbs is that low-income and minority people are unable to live in them because of the high cost of the housing and the unfriendliness of established residents. Legal devices, such as requiring each house to sit on a large lot and the prohibition of apartments, prevent low-income families from living in many suburbs.

Contribution of Transportation to Suburbanization
Urban sprawl makes people more dependent on transportation for access to work, shopping, and leisure activities. More than half of all trips are work-related. Shopping or other personal business and social journeys each account for approximately one-fourth of all trips. Historically, the growth of suburbs was constrained by transportation problems. People lived in crowded cities

because they had to be within walking distance of shops and places of employment. Cities then built street railways . . . and underground railways. Many so-called streetcar suburbs built in the nineteenth century still exist and retain unique visual identities.

Motor Vehicles

The suburban explosion in the twentieth century has relied on motor vehicles rather than railroads, especially in the United States. Rail and trolley lines restricted suburban development to narrow ribbons within walking distance of the stations. Motor vehicle ownership is nearly universal among American households. Outside the big cities, public transportation service is extremely rare or nonexistent. The U.S. government has encouraged the use of cars and trucks by paying 90 percent of the cost of limited-access high-speed interstate highways . . . (and) by policies that limit the price of fuel to less than one half the level found in Western Europe.

(463)

The motor vehicle is an important user of land in the city. An average city allocates about one-fourth of its land to roads and parking lots. European and Japanese cities have been especially disrupted by attempts to insert new roads and parking areas in or near to the medieval central areas. Technological improvements may help traffic flow. Computers mounted on the dashboards alert drivers to traffic jams and suggest alternate routes. On freeways, vehicle speed and separation from other vehicles can be controlled automatically. The inevitable diffusion of such technology in the twenty-first century will reflect the continuing preference of most people in MDCs to use private motor vehicles rather than switch to public transportation.

Public Transportation

Because few people in the United States live within walking distance of their place of employment, urban areas are characterized by extensive commuting.

Rush-Hour Commuting. As much as 40 percent of all trips made into or out of a CBD occur during four hours of the day—two in the morning and two in the afternoon. **Rush hour**, or peak hour, is the four consecutive 15-minute periods that have the heaviest traffic. Public transportation is better suited than motor vehicles to moving large numbers of people. But most Americans still prefer to commute by car. Public transportation is cheaper, less polluting, and more energy-efficient than the automobile. Its use is increasingly confined in the United States to rush-hour commuting by workers in the CBD.

(464)

Automobiles have costs beyond their purchase and operation: delays imposed on others, increased need for highway maintenance, construction of new highways, and pollution. Yet despite the obvious advantages of public transportation for commuting, ridership in the United States declined from 23 billion per year in the 1940s to 8 billion in 2002. The number of U.S. and Canadian cities with trolley service declined from approximately 50 in 1950 to 8 in the 1960s. General Motors acquired many of the privately owned streetcar companies and replaced the trolleys with buses that the company made. Bus ridership declined from a peak of 11 billion riders annually in the late 1940s to 6 billion in 2001. Commuter railroad service, like trolleys and buses, has also been drastically reduced in most U.S. cities.

New Rapid Transit Lines. The one exception to the downward trend in public transportation is rapid transit.

(465)

Cities such as Boston and Chicago have attracted new passengers through construction of new subway lines and modernization of existing service. Entirely new subway systems have been built in recent years in U.S. cities, including Atlanta, Baltimore, Miami, San Francisco, and Washington, D.C. The federal government has permitted Boston, New York, and other cities to use funds originally allocated for interstate highways to modernize rapid transit service instead. Subway ridership in the United States has increased 2 percent each year since 1980. The trolley—now known (as) . . . fixed light-rail transit—is making a modest comeback in North

America. However, new construction in all 10 cities amounted only to about 200 kilometers (130 miles) since 1980, and ridership in all cities combined is 1 million a day.

California, the state that most symbolizes the automobile-oriented American culture, leads in construction of new fixed light-rail transit lines. Los Angeles—the city perhaps most associated with the motor vehicle—has planned the most extensive new light-rail system . . . but construction is very expensive, and the lines (will) serve only a tiny percentage of the region.

Service Versus Cost. Low-income people tend to live in inner-city neighborhoods, but the job opportunities . . . are in suburban areas not well served by public transportation. Despite modest recent successes, most public transportation systems are caught in a vicious circle, because fares do not cover operating costs. As patronage declines and expenses rise, the fares are increased, which drives away passengers and leads to service reduction and still higher fares. The United States does not fully recognize that public transportation is a vital utility deserving of subsidy to the degree long assumed by European governments.

Public Transit in Other Countries. In contrast, even in more developed Western European countries and Japan, where automobile ownership rates are high, extensive networks of bus, tram, and subway lines have been maintained, and funds for new construction have been provided in recent years.

(466)
Local Government Fragmentation
The fragmentation of local government in the United States makes it difficult to solve regional problems of traffic, solid-waste disposal, and construction of affordable housing.

(467)
Metropolitan Government
The large number of local government units has led to calls for a metropolitan government that could coordinate—if not replace—the numerous local governments in an urban area. Most U.S. metropolitan areas have a **council of government**, which is a cooperative agency consisting of representatives of the various local governments in the region. Strong metropolitan-wide governments have been established in a few places in North America. Two kinds exist: federations and consolidations.

Federations. Toronto, Ontario, has a federation system. The region's six local governments . . . are responsible for police, fire, and tax-collection services. A regional government, known as the Metropolitan Council, or Metro, sets the tax rate . . . borrows money for new projects. Metro shares responsibility with local governments for public services, such as transportation, planning, parks, water, sewage, and welfare.

Consolidations. Several U.S. urban areas have consolidated metropolitan governments; Indianapolis and Miami are examples. Both have consolidated city and county governments.

Growing Smart
Several U.S. states have taken strong steps in the past few years to curb sprawl, reduce traffic congestion, and reverse inner-city decline. Legislation and regulations to limit suburban sprawl and preserve farmland has been called **smart growth**. Maryland enacted especially strong smart growth legislation in 1998. State money must be spent to "fill in" already urbanized areas. Oregon and Tennessee have defined growth boundaries within which new development must occur. New Jersey, Rhode Island, and Washington were also early leaders in enacting strong state-level smart-growth initiatives.

Key Terms
Annexation (p.458)

Census tract (p.446)

Concentric zone model (p.445)

Council of government (p.467)

Density gradient (p.460)

Edge city (p.460)

Filtering (p.455)

Gentrification (p.456)

Chapter 14. Resource Issues

People have always transformed Earth's land, water, and air for their benefit. But human actions in recent years have gone far beyond the impact of the past. The magnitude of transformations is disproportionately shared by North Americans; with only one-twentieth of Earth's population, North Americans consume one-fourth of the world's energy and generate one-fourth of many pollutants. Elsewhere in the world, two billion people live without clean water or sewers. One billion live in cities with unsafe sulfur dioxide levels. Future generations will pay the price if we continue to mismanage Earth's resources. Our carelessness has already led to unsafe drinking water and toxic air in some places. Our inefficiency could lead to shortages of food. Humans once believed Earth's resources to be infinite, or at least so vast that human actions could never harm or deplete them. But warnings from scientists, geographers, and governments are making clear that resources are indeed a problem.

Key Issues
1. Why are resources being depleted?
2. Why are resources being polluted?
3. Why are resources reusable?
4. Why can resources be conserved?

(475)
Geographers study the troubled relationship between human actions and the physical environment in which we live. A resource is a substance in the environment that is useful to people, is economically and technologically feasible to access, and is socially acceptable to use. Resources include food, water, soil, plants, animals, and minerals. The problem is that most resources are limited, and Earth has a tremendous number of consumers. Geographers observe two major misuses of resources:
1. We deplete scarce resources—especially petroleum, natural gas, and coal . . .
2. We destroy resources through pollution of air, water, and soil.

Nowhere is the globalization trend more pronounced than in the study of resources. Global uniformity in cultural preferences means that people in different places value similar natural resources, although not everyone has the same access to them. In a global environment, all places are connected, so the misuse of a resource in one place affects the well-being of people everywhere.

Key Issue 1. Why Are Fossil-Fuel Resources Being Depleted?
- **Energy Resources**
- **Mineral Resources**

Natural resources have little value in and of themselves. Their value derives from their usefulness to humans, especially in production. Two kinds of natural resources are especially valuable to humans: minerals and energy resources. More developed countries want to preserve current standards of living, and less developed countries are struggling to attain a better standard. All this demands tremendous energy resources, so as we deplete our current sources of energy, we must develop alternative ones.

Energy Resources
Historically, people relied on power supplied by themselves or by animals, known as animate power. Ever since the Industrial Revolution began in the late 1700s, humans have expanded their use of inanimate power, generated from machines. Humans have found the technology to harness the great potential energy stored in resources such as coal, oil, gas, and uranium.

Three of Earth's substances provide five-sixths of the world's energy: oil, natural gas, and coal. In MDCs the remainder comes primarily from nuclear, solar, hydroelectric, and geothermal power. Burning wood provides much of the remaining energy in less developed societies. Historically, the most important energy source worldwide was biomass fuel, such as wood, plant material, and animal waste. Coal supplanted wood as the leading energy source in the late 1800s in North America and

Western Europe. Petroleum was . . . not an important resource until the diffusion of automobiles. Natural gas was originally burned off as a waste product of oil drilling but now heats millions of homes. Energy is used in three principal places: businesses, homes, and transportation. For U.S. businesses the main energy resource is coal. Some businesses directly burn coal. Others rely on electricity, mostly generated at coal-burning power plants. At home, energy is used primarily for heating of living space and water. Natural gas is the most common source, followed by petroleum (heating oil and kerosene).

(476)
Almost all transportation systems operate on petroleum products. Only subways, streetcars, and some trains run on coal-generated electricity. Petroleum, natural gas, and coal are known as fossil fuels. A fossil fuel is the residue of plants and animals that became buried millions of years ago.

Two characteristics of fossil fuels cause great concern for the future:
1. The supply of fossil fuels is finite;
2. Fossil fuels are distributed unevenly around the globe.

Finiteness of Fossil Fuels
To understand Earth's resources, we distinguish between those that are renewable and those that are not:
- Renewable energy is replaced continually, or at least within a human lifespan;
- Nonrenewable energy forms so slowly that . . . it cannot be renewed.

The world faces an energy problem in part because we are rapidly depleting the remaining supply of the three fossil fuels, especially petroleum. We can use other resources for heat, fuel, and manufacturing, but they are likely to be more expensive and less convenient to use than fossil fuels.

Remaining Supply of Fossil Fuels. How much of the fossil-fuel supply remains? Despite the critical importance of this question for the future, no one can answer it precisely. The amount of energy remaining in deposits that have been discovered is called a proven reserve. Proven reserves can be measured with reasonable accuracy.

(477)
The energy in undiscovered deposits that are thought to exist is a potential reserve. The World Energy Council estimates potential oil reserves of about 500 billion barrels, with the largest fields thought to lie beneath the South China Sea and northwestern China. At the current world petroleum consumption rate of about 25 billion barrels a year, Earth's proven petroleum reserves of 1 trillion barrels will last 40 years.

Petroleum is being consumed at a more rapid rate than it is being found, and world demand is increasing by more than 1 percent annually. Similarly, at current rates of use, the world's proven reserves of natural gas will last for about 80 years . . . because the world currently uses much more oil than gas. At current consumption, proven coal reserves can last at least several hundred years. More than one-half of U.S. electricity currently comes from power plants that burn coal.

Extraction of Remaining Reserves. When it was first exploited, petroleum "gushed" from wells. . . . Coal was quarried in open pits. But now extraction is harder. The largest, most accessible deposits of (fossil fuels) already have been exploited. Newly discovered reserves generally are smaller and more remote. Unconventional sources of petroleum and natural gas are being studied and developed, such as oil shale and tar sandstones. They are called unconventional . . . because we do not currently have economically feasible, environmentally sound technology to extract them. Utah, Wyoming, and Colorado contain more than 10 times the petroleum reserves of Saudi Arabia, but as oil shale.

Uneven Distribution of Fossil Fuels
Geographers observe two important inequalities in the global distribution of fossil fuels:

- Some regions have abundant reserves, whereas others have little;
- Consumption of fossil fuels is much higher in some regions than in others.

Unequal possession and consumption of fossil fuels have been major sources of global instability in the world.

Location of Reserves. Some regions have abundant reserves of fossil fuels, but other regions have little. This partly reflects how fossil fuels form. Coal forms in tropical locations. The tropical swamps of 250 million years ago have relocated to the mid latitudes . . . (with) the slow movement of Earth's drifting continents.

(478)
Oil and natural gas formed . . . from sediment deposited on the seafloor. Some . . . reserves still lie beneath seas . . . , but other reserves are located beneath land that had been under water millions of years ago. . . . Five Middle Eastern countries have two-thirds of the world's oil reserves. Venezuela and Mexico have the most extensive proven reserves in the Western Hemisphere. The United States accounts for 10 percent of the world's annual production of petroleum but possesses only 2 percent of the proven reserves. Russia accounts for one-fourth of world production of natural gas and possesses one-third of the world's proven reserves. The United States also produces one-fourth of the world's natural gas but has only 3 percent of the world's reserves.

A handful of less developed countries in Africa, Asia, and Latin America have extensive reserves of one or another of the fossil fuels, but most have little. More developed countries historically have possessed a disproportionately high percentage of the world's fossil-fuel reserves. The MDCs produced most of the world's fossil fuels during the nineteenth and twentieth centuries. But this dominance is likely to end in the twenty-first century.

Most of the world's proven reserves (and probably potential reserves) are in a handful of Asian countries, especially China, the Middle East, and former Soviet Union republics.

(479)
Consumption of Fossil Fuels. The global pattern of fossil-fuel consumption—like production—will shift in the twenty-first century. More developed countries, with about one-fourth of the world's population, currently consume about three-fourths of the world's energy. The sharp regional difference in energy consumption has two geographic consequences for the future:
- As they promote development and cope with high population growth, LDCs will consume much more energy. The share of world energy consumed by LDCs will increase from about 25 percent today to 40 percent by 2010 and 60 percent by 2020.
- MDCs . . . must import fossil fuels . . . from LDCs. Because of development and population growth in LDCs, the more developed countries will face greater competition in obtaining the world's remaining supplies of fossil fuels.

Control of World Petroleum
The sharpest conflicts over energy will be centered on the world's limited proven reserves of petroleum.

(480)
The MDCs import most of their petroleum from the Middle East, where most of the world's proven reserves are concentrated. Both U.S. and Western European transnational companies originally exploited Middle Eastern petroleum fields and sold the petroleum at a low price to consumers in MDCs.

OPEC Policies during the 1970s. Several LDCs possessing substantial petroleum reserves created the Organization of Petroleum Exporting Countries (OPEC) in 1960. OPEC's Arab members were angry at North American and Western European countries for supporting Israel during that nation's

1973 war with the Arab states of Egypt, Jordan, and Syria. So during the winter of 1973–74, they flexed their new economic muscle with a boycott—Arab OPEC states refused to sell petroleum to the nations that had supported Israel. Soon gasoline supplies dwindled in MDCs. Each U.S. gasoline station was rationed a small quantity of fuel. European countries took more drastic action—the Netherlands, for example, banned all but emergency motor vehicle travel on Sundays.

OPEC lifted the boycott in 1974 but raised petroleum prices from $3 per barrel to more than $35 by 1981. The rapid escalation in petroleum prices caused severe economic problems in MDCs during the 1970s.

(481)
Many manufacturers were forced out of business. . . . The LDCs were hurt even more. They depended on low-cost petroleum imports to spur industrial growth. Their fertilizer costs shot up, because many fertilizers are derived from oil. North American and Western European states . . . encouraged OPEC countries to invest in American and European real estate, banks, and other safe and profitable investments.

(482)
Comparable investment opportunities were limited in less developed countries.

Reduced Influence of OPEC. Internal conflicts weakened OPEC's influence in the 1980s and 1990s. By not acting together, individual OPEC members produced more petroleum than the world demanded, and MDCs stockpiled some of the surplus as protection against another boycott. Conservation measures dampened demand for petroleum in most developed countries. The fuel efficiency of the average new car sold in the United States increased from 16 miles per gallon in 1975 to 27 miles per gallon in 2000. However, the number of vehicles . . . doubled . . . and the average vehicle was driven 10 percent more. Further, savings from more efficient cars have been offset by increased sales of gas-guzzling sport-utility vehicles and pickup trucks.
The United States reduced its dependency on imported oil in the immediate wake of the 1970s shocks, and the share of imports from OPEC countries declined from two-thirds in 1973 to one-half in 1999. But oil imports climbed rapidly in the late twentieth century, from 1.2 billion barrels in 1985 to 3.2 billion barrels in 1999. At some point extracting the remaining petroleum reserves will prove so expensive and environmentally damaging that use of alternative energy sources will accelerate, and dependency on petroleum will diminish. The issues for the world are whether dwindling petroleum reserves are handled wisely and other energy sources are substituted peacefully. Given the massive growth . . . in LDCs . . . more developed countries may have little influence. . . .

Nonrenewable Substitutes for Petroleum
As petroleum supplies dwindle, the two other principal fossil fuels—natural gas and coal—are short-run substitutes. Nuclear energy also figures prominently in short-term energy planning.

Natural Gas. Natural gas is cheaper to burn and is less polluting than petroleum and coal, and in the twentieth century supplies were less subject to disruptions for political reasons. Consequently, world natural gas consumption increased by two-thirds between 1980 and 2000, whereas petroleum consumption increased by only one-fifth. At the current rate of use, the world's proven reserves of about 140 trillion cubic meters of natural gas will last for about 80 years. Although the United States is a major producer of natural gas, proven reserves are limited.

(483)
Within North America, pipelines carry natural gas to industrial and residential users. It is difficult to ship natural gas across oceans.

Coal. At current consumption, the world's proven coal reserves of one quadrillion metric tons can last several hundred years. But problems hinder expanded use of coal: air pollution, mine safety, land subsidence, and economics.

Uncontrolled burning of coal releases several pollutants. The U.S. Clean Air Act now requires utilities to use better-quality coal or to install "scrubbers" on smokestacks.

Historically, mining was an especially dangerous occupation. Strictly enforced U.S. mine safety laws, . . . automation of mining, and a smaller workforce have made the American coal industry much safer. Annual U.S. mine mortality now is below 100. But that figure could rise if mining operations expanded.

Underground mining may . . . pollute water . . . and (cause) subsidence or sinking of the ground. . . . Surface mining can cause soil erosion. In the United States the mining industry is highly regulated, and most companies today have a good record. . . . But . . . practices in the past have left a legacy of environmental damage.

Coal must be shipped long distances, because most of the factories and power plants using it are far from the coalfields. Ironically, the principal methods of transporting coal—barge, rail, or truck—are all powered by petroleum.

Nuclear Energy. The big advantage of nuclear power is the large amount of energy that is released from a small amount of material. Nuclear power supplies about one-sixth of the world's electricity. The United States is responsible for generating one-third of the world's nuclear power, France and Japan together another one-third. About 30 countries make some use of nuclear power.
The countries most highly dependent on nuclear power are clustered in Europe. Lithuania and France obtain more than three-fourths of their electricity from nuclear power, Belgium, Bulgaria, Slovakia, and Ukraine about one-half each. About one-fifth of electricity in the United States comes from nuclear power, about one-third in Japan.

Dependency on nuclear power varies widely among U.S. states. Nuclear power provides one half of the electricity in New England, one fourth in the Southeast and the Midwest, and only one-tenth in states west of the Mississippi River. Nuclear power presents serious problems. These include potential accidents, radioactive waste, generation of plutonium, a limited uranium supply, geographic distribution, and cost.

1. Potential Accidents. A nuclear power plant produces electricity from energy released by splitting uranium atoms in a controlled environment, a process called fission.

(484)
Elaborate safety precautions are taken to prevent nuclear fuel from leaking from a power plant. Nuclear power plants cannot explode, like a nuclear bomb. . . . However, it is possible to have a runaway reaction, which overheats the reactor, causing a meltdown, possible steam explosions, and scattering of radioactive material into the atmosphere. This happened in 1986 at Chernobyl . . . in the north of Ukraine, near the Belarus border.

(485)
The impact of the Chernobyl accident extended through Europe. Half of the eventual victims may be residents of European countries other than Ukraine and Belarus.

2. Radioactive Waste. When nuclear fuel fissions, the waste is highly radioactive and lethal and remains so for many years. Plutonium can be harvested from it for making nuclear weapons. No one has yet devised permanent storage for radioactive waste. It must be isolated for several thousand years. The United States is Earth's third-largest country in land area, yet it has failed to find a suitable underground storage site because of worry about groundwater contamination. The time required for radioactive waste to decay to a safe level is far longer than any country or civilization has existed.

3. Bomb Material. Nuclear power has been used in warfare twice, in August 1945, when the United States dropped an atomic bomb on first Hiroshima and then Nagasaki, Japan. No government has (since) dared to use them in a war, because leaders have recognized that a full-scale nuclear conflict could terminate human civilization. But the black market could provide terrorists with enough plutonium to construct nuclear weapons.

4. Limited Uranium Reserves. Like fossil fuels, proven uranium reserves are limited—about 60 years at current rates of use. Uranium ore naturally contains only 0.7 percent U-235, and a greater concentration is needed for power generation. Proven uranium reserves could be depleted in three more decades. A breeder reactor turns uranium into a renewable resource by generating plutonium, also a nuclear fuel. However, plutonium is more lethal than uranium. It is also easier to fashion into a bomb.

(486)
Because of these risks, few breeder reactors have been built, and none are in the United States.
5. Cost. Nuclear power plants cost several billion dollars to build, primarily because of elaborate safety measures. Uranium is mined in one place, refined in another, and used in still another. The cost of generating electricity is much higher from nuclear plants than from coal-burning plants. The future of nuclear power has been seriously hurt by the combination of high risk and cost. Most countries in North America and Western Europe have curtailed construction of new plants. Nuclear power will decline in other countries as older nuclear plants are closed and not replaced.

Mineral Resources
Earth has 92 natural elements, but about 99 percent of the crust is composed of 8 elements: oxygen, silicon, aluminum, iron, calcium, sodium, potassium, and magnesium. The eight most common elements combine with . . . rare ones to form approximately 3,000 different minerals, all with their own properties of hardness, color, and density, as well as spatial distribution. Each mineral potentially is a resource, if people find a use for it.

When a new technological process or product is invented, demand can suddenly increase for a mineral that had little use in the past. Conversely, when a new process or product replaces an older one, demand may decline for a mineral important in the past. Mineral deposits are not uniformly distributed around the world. A handful of countries accounts for most of the world's supply of particular minerals. Further, the leading producers at this time are not always the countries with the most extensive reserves, an indication that the relative fortunes of states may change in the future.

In weight, more than 90 percent of the minerals that humans use are nonmetallic, but metallic minerals are important for economic activities and so carry relatively high value.

Nonmetallic Minerals
Building stones, including large stones, coarse gravel, and fine sand, account for 90 percent of nonmetallic mineral extraction. Nonmetallic minerals are also used for fertilizer. Important nonmetallic mineral sources of fertilizers include phosphorus, potassium, calcium, and sulfur. All four are abundant elements in nature with wide distributions. However, mining is highly clustered where the minerals are most easily and cheaply extracted.

(487)
Although only a small percentage of nonmetallic minerals in weight, gemstones are valued especially highly for their color and brilliance when cut and polished. Diamonds are especially useful in manufacturing.

Metallic Minerals
Metallic minerals have properties that are especially valuable for fashioning machinery, vehicles, and other essential components of an industrialized society. Many metals are also capable of combining with other metals to form alloys with yet other distinctive properties.

Metals are known as ferrous or nonferrous:
- Ferrous is derived from the Latin word for iron;
- Nonferrous metals are those utilized to make products other than iron and steel.

World supply of most metals is high, including the most widely used ferrous metal (iron) and the most widely used nonferrous metal (aluminum). However, reserves of some metals are low, posing a challenge to manufacturers to find economically feasible substitutes.

Ferrous Metals. By far the world's most widely used ferrous metal is iron, which accounts for 5 percent of Earth's crust by weight and 95 percent of ferrous metal mineral extraction.

The critical importance of iron to the past four thousand years of human history is reflected by the application of the term "Iron Age" to the period. Mining of iron ore, from which iron is extracted, is concentrated in a handful of countries.

(488)
Iron deposits of indifferent quality but close to market are actively mined, whereas large known deposits in remote areas are ignored for now, although they may become more important in the future once more accessible deposits are exhausted.

Other less common ferrous metals are important for alloying with iron to produce steel. Important alloying elements in abundant supply include manganese, chromium, titanium, magnesium, and molybdenum. Supplies of other alloying elements, notably nickel, tin, and tungsten, are limited.

Nonferrous Metals. Rarely used commercially prior to the twentieth century, aluminum is now in greater demand than any metal except iron. World supply of aluminum is so large—more than 1,000 years at current rates of use—that it is essentially regarded as inexhaustible at realistic projections of future demand.

Three other especially important nonferrous elements are copper, lead, and zinc.

(489)
World supplies are extremely limited—less than 60 years for copper, 25 years for lead, and 45 years for zinc.

Nonferrous metals also include precious metals: silver, gold, and the platinum group.
Silver and gold have been prized since ancient times for their beauty and durability.
The platinum group includes six related metals that commonly occur together in nature and are especially scarce: platinum, palladium, rhodium, ruthenium, iridium, and osmium. The principal use of the platinum group is in motor-vehicle catalytic converters to treat exhaust emissions, as well as fuel cells.

Key Issue 2. Why Are Resources Being Polluted?
- **Air pollution**
- **Water pollution**
- **Land pollution**

In our consideration of resources, consumption is half of the equation—waste disposal is the other half. We rely on air, water, and land to remove and disperse our waste. When more waste is added than a resource can accommodate, we have pollution. Natural processes may transport pollutants from one part of the environment to another. Discharges to the air often turn up in rivers, and wastes dumped in landfills can produce gases that leak into the atmosphere. Not all pollution is caused by humans. However, our focus here is on the pollution that humans cause.

(490)
Air Pollution. Air pollution is a concentration of trace substances at a greater level than occurs in average air. The most common air pollutants are carbon monoxide, sulfur dioxide, nitrogen oxides, hydrocarbons, and solid particulates. Three human activities generate most air pollution: motor vehicles, industry, and power plants. In all three cases, pollution results from the burning of fossil fuels. Air pollution concerns geographers at three scales—global, regional, and local.

Global Scale Air Pollution
Air pollution may contribute to global warming and damage the atmosphere's ozone layer. It may also be damaging the atmosphere's ozone layer.

Global Warming. Human actions, especially the burning of fossil fuels, may be causing Earth's temperature to rise. Earth is warmed by sunlight that . . . is converted to heat. When the heat tries to pass back through the atmosphere to space, some gets through and some is trapped. A concentration of trace gases in the atmosphere can block or delay the return of some of the heat leaving the surface heading for space. Plants and oceans absorb much of the discharges, but increased fossil-fuel burning during the past 200 years has caused the level of carbon dioxide in the atmosphere to rise by more than one-fourth. The average temperature of Earth's surface has increased by 1° Celsius (2° Fahrenheit) during the past century. The buildup of carbon dioxide contributed to the warming, although scientists disagree on whether it caused most or only a small percentage of the warming. The anticipated increase in Earth's temperature, caused by carbon dioxide trapping some of the radiation emitted by the surface, is called the greenhouse effect.

(491)
Global warming of only a few degrees could melt the polar ice caps and raise the level of the oceans many meters. Global patterns of precipitation could shift. The shifts in coastlines and precipitation patterns could require massive migration and be accompanied by political disputes.

Global-Scale Ozone Damage. The stratosphere . . . contains a concentration of ozone gas. The ozone layer absorbs dangerous ultraviolet (UV) rays from the Sun. Earth's protective ozone layer is threatened by pollutants called chlorofluorocarbons (CFCs). The 1987 Montreal Protocol called for more developed countries to cease using CFCs by 2000, and for LDCs to cease by 2010.

Regional Scale Air Pollution
Air pollution may damage a region's vegetation and water supply through acid deposition. Industrialized, densely populated regions in Europe and eastern North America are especially affected by acid deposition. Sulfur oxides and nitrogen oxides, emitted by burning fossil fuels, enter the atmosphere, where they combine with oxygen and water. Tiny droplets of sulfuric acid and nitric acid form and return to Earth's surface as acid deposition. When dissolved in water, the acids may fall as acid precipitation—rain, snow, or fog. These acidic droplets might be carried hundreds of kilometers. Acid precipitation has damaged lakes, killing fish and plants. On land, concentrations of acid in the soil can injure plants by depriving them of nutrients and can harm soil worms and insects. Buildings and monuments made of marble and limestone have suffered corrosion from acid rain. Despite progress (in reducing sulfur dioxide emissions), acid precipitation continues to damage forests and lakes. Governments are reluctant to impose the high cost of controls on their industries and consumers.

(492)
Geographers are particularly interested in the effects of acid precipitation because the worst damage is not experienced at the same location as the emission of the pollutants.
(493)
Acid rain falling in Ontario, Canada, for example, can be traced to emissions from coal-burning power plants in the U.S. Great Lakes. Eastern Europe has suffered especially severely from acid precipitation, a legacy of Communist policies that encouraged the construction of factories and power plants without pollution-control devices.

172

The destruction of trees has harmed Eastern Europe's seasonal water flow. In dense forests, snow used to melt slowly and trickle into rivers. Now, on the barren sites, it melts and drains quickly, causing flooding in the spring and water shortages in the summer. Perhaps the most severe impact is on human life. A 40-year-old man living in Poland's polluted southern industrial area has a life expectancy 10 years less than his father had at the same age.

Local Scale Air Pollution

Air pollution is especially severe in local areas where emission sources are concentrated, such as urban areas. Urban areas may be especially polluted because . . . polluters emit residuals in a concentrated area. Urban air pollution has three basic components:

1. **Carbon monoxide**;
2. **Hydrocarbons**—hydrocarbons and nitrogen oxides in the presence of sunlight form photochemical smog;
3. **Particulates**—particulates include dust, smoke, (and rubber tire) particles.

The severity of air pollution . . . depends on the weather. The worst urban air pollution occurs when winds are slight, skies are clear, and a temperature inversion exists. The worst U.S. city for concentrations of carbon monoxide and second worst for particulates is Denver. The Rocky Mountains help trap the gases and produce a permanent temperature inversion. The problem is not confined to MDCs. Santiago, Chile, nestled between the Pacific Ocean and the Andes Mountains, suffers severe smog problems.

(494)

Progress in controlling urban air pollution is mixed. Air has improved in developed countries where strict clean-air regulations are enforced. But more people are driving, offsetting gains made by emission controls. Limited emission controls in LDCs are contributing to severe urban air pollution.

Water Pollution

Water serves many human purposes. These uses depend on fresh, clean, unpolluted water. Clean water is not always available, because people also use water for purposes that pollute it. Pollution is widespread, because it is easy to dump waste into a river and let the water carry it downstream where it becomes someone else's problem.

(495)

Water Pollution Sources

Three main sources generate most water pollution:

- **Water-Using Industries**: Industries . . . generate a lot of wastewater. Water can also be polluted by industrial accidents.
- **Municipal Sewage**: In more developed countries, sewers carry wastewater . . . to a municipal treatment plant, where most—but not all—of the pollutants are removed. In LDCs, sewer systems are rare, and wastewater usually drains untreated into rivers and lakes.
- **Agriculture**: Fertilizers and pesticides spread on fields to increase agricultural productivity are carried into rivers and lakes by the irrigation system or natural runoff.

Point-source pollution enters a stream at a specific location, whereas nonpoint-source pollution comes from a large diffuse area. Farmers tend to pollute through nonpoint sources, such as by permitting fertilizer to wash from a field during a storm. Point-source pollutants are usually smaller in quantity and much easier to control.

(496)

Impact on Aquatic Life

If too much waste is discharged into the water, the water becomes oxygen-starved and fish die. The oxygen consumed by the decomposing organic waste constitutes the biochemical oxygen demand (BOD). When runoff carries fertilizer from farm fields into streams or lakes, the fertilizer nourishes excessive aquatic plant production . . . that consumes too much oxygen. Either type of pollution unbalances the normal oxygen level, threatening aquatic plants and animals. Some of the residuals may become concentrated in the fish, making them unsafe for human consumption. Many factories and power plants use water for cooling and then discharge the warm water back into the river or lake. Fish adapted to cold water, such as salmon and trout, might not be able to survive in the warmer water.

Wastewater and Disease
Since passage of . . . clean water . . . laws . . . most treatment plants meet high water-quality standards. Improved treatment procedures have resulted in cleaner rivers and lakes in more developed countries. Prior to the Industrial Revolution, the Thames was a major food source for Londoners. During the Industrial Revolution, the Thames became the principal location for dumping waste. The fish died, and the water grew unsafe to drink. The British government began a massive cleanup during the 1960s to restore the Thames to health. A salmon was caught in the Thames just upstream from London in 1982, the first since 1833. In LDCs, sewage often flows untreated directly into rivers. The drinking water, usually removed from the same rivers, may be inadequately treated as well. Waterborne diseases such as cholera, typhoid, and dysentery are major causes of death. In less developed countries . . . industrialization may take a higher priority than clean water.

Land Pollution
When we consume a product, we also consume an unwanted byproduct . . . (the) container in which the product is packaged. About 2 kilograms (4 pounds) of solid waste per person is generated daily in the United States.

(497)
Even consumers who carefully dispose of solid waste are contributing to a major pollution problem. About one-half of the solid waste generated in the United States is placed in landfills, the other one half recycled in one of two ways. Most of the recycled solid waste is reused, while the remainder is incinerated for power.

Solid Waste Disposal
The sanitary landfill is by far the most common strategy for disposal of solid waste in the United States. We disperse air and water pollutants into the atmosphere, rivers, and eventually the ocean, but we concentrate solid waste in thousands of landfills. Chemicals released by the decomposing solid waste can leak from the landfill into groundwater. This can contaminate water wells, soil, and nearby streams. Eventually, landfills fill up. Few new ones are being built.

Incineration. Burning the trash reduces its bulk by about three-fourths, and the remaining ash demands far less landfill space. Incineration also provides energy . . . to produce steam heat or to operate a turbine that generates electricity. The percentage of solid waste that is burned has increased rapidly during the past quarter century. However, solid waste, a mixture of many materials, may burn inefficiently. Burning releases some toxics into the air, and some remain in the ash.

Toxic Pollutants
Disposing of toxic wastes is especially difficult. Toxic wastes include heavy metals (including mercury, cadmium, and zinc), PCB oils from electrical equipment, cyanides, strong solvents, acids, and caustics. Burial of wastes was once believed to be sufficient to handle the disposal problem, but many of the burial sites have leaked. One of the most notorious is Love Canal, near Niagara Falls, New York. Love Canal is not unique. Toxic wastes have been improperly disposed of at thousands of dumps. Companies in the United States that release chemicals classified as toxic by the E.P.A. must report the amounts released. About 2.5 billion kilograms (6 billion pounds) of toxic chemicals are discharged in the United States.

(498)
About one-fourth of the discharges are by 10 companies. Some European and North American firms have tried to transport their waste to West Africa, often unscrupulously.

Key Issue 3. Why Are Resources Reusable?
- **Renewing resources**
- **Recycling resources**

Depletion and destruction of resources can be reduced through reusing resources. Nonrenewable resources can be replaced with renewable ones. Discharging of unwanted byproducts into the environment can be replaced with recycling of them into resources.

Renewing Resources
Energy poses an especially strong challenge in substituting renewable resources for nonrenewable ones. About 6 percent of energy consumed in the United States is generated by renewable sources.

Solar Energy
Solar energy is free, does not damage the environment or cause pollution, and is quite safe. There are two general approaches to harnessing solar energy: passive and active.

Passive solar energy. Passive solar energy systems capture energy without special devices. Reliance on passive solar energy increased during the nineteenth century when construction innovations first permitted hanging of massive glass "curtains " on a thin steel frame.

With electricity and petroleum cheap and abundant, passive solar energy did not play a major role in construction of homes and commercial buildings through most of the twentieth century. Interest in passive solar energy resumed when petroleum prices rapidly escalated during the 1970s. The largest contributor to increased use of passive solar energy has been through advances in glass technology.

Active solar energy. Active solar energy systems collect solar energy and convert it either to heat energy or to electricity. In direct electric conversion, solar radiation is captured with photovoltaic cells. In indirect electric conversion, solar radiation is first converted to heat, then to electricity. In heat conversion, solar radiation is concentrated with large reflectors and lenses to heat water or rocks.

(499)
Other Energy Sources
Other energy sources include hydroelectric, geothermal, biomass, and fusion. The first three are currently used but offer limited prospects for expansion. Fusion is not a practical source at this time but may be in the future.

Hydroelectric Power. Water has been a source of mechanical power since before recorded history. It turned water wheels . . . to operate machines. Over the last hundred years the energy of moving water has been used to generate electricity, called hydroelectric power.
Hydroelectric power is the world's second most popular source of electricity, after coal, supplying about a fourth of worldwide demand. Hydroelectric power has drawbacks. Dams may flood formerly usable land, cause erosion, and upset ecosystems. Political problems can result from building dams on rivers that flow through more than one country.

Geothermal Energy. Earth's interior is hot from natural nuclear reactions. In volcanic areas . . . hot rocks can encounter groundwater. Energy from this hot water or steam is called geothermal energy. Harnessing geothermal energy is most feasible at the rifts along Earth's surface where crustal plates meet.

Biomass. Forms of biomass, such as sugarcane, corn, and soybeans, can be processed into motor vehicle fuels. Burning biomass may be inefficient, because the energy used to produce the crops may be as much as the energy supplied by the crops. The most important limitation on using biomass for energy is that it already serves other essential purposes: providing much of Earth's food, clothing, and shelter.

Nuclear Fusion. Some nuclear power problems could be solved with nuclear fusion, which is the fusing of hydrogen atoms to form helium. But fusion can occur only at very high temperatures . . . (which have not been) achieved . . . on a sustained basis in a power-plant reactor. Alternatives such as fusion do not offer immediate solutions to energy shortages in the twenty-first century. The era of dependency on nonrenewable fossil fuels for energy will constitute a remarkably short period of human history.

Uses for Renewable Energy
Efforts to utilize more renewable energy are focused on two sectors: electricity and motor vehicles.

Electricity. In MDCs, solar energy is used primarily as a substitute for electricity in heating water. Solar-generated electricity is used in spacecraft, light-powered calculators, and at remote sites where conventional power is unavailable, such as California's Mojave Desert. The largest and fastest growing market for photovoltaic cells are the two billion people who lack electricity in LDCs, especially residents of remote villages. The cost of cells must drop and their efficiency must improve for solar power to expand rapidly, with or without government support.

(500)
Motor Vehicles. The most serious obstacle to decreasing reliance on nonrenewable energy is its importance as automotive fuel. The use of electric vehicles is expanding in MDCs, primarily to reduce air pollution rather than to conserve the nonrenewable resource of petroleum. Limitations with electric power have led motor vehicle producers to consider fuel cells instead. As long as petroleum is perceived as cheap and unlimited in supply, alternative fuel vehicles have limited popularity. It will take a major increase in world petroleum prices or disruption in supplies to bring alternative fuel vehicles to the market in large numbers.

Recycling Resources
Recycling increased in the United States from 7 percent of all solid waste in 1970 to 10 percent in 1980, 17 percent in 1990, and 28 percent in 1999. The amount of solid waste generated by Americans increased by 30 million tons during the 1990s, and all of that additional waste was recycled. The percentage of recovered materials varies widely by product.

Recycling Collection
Recycling involves two main series of activities. First, materials that would otherwise be "thrown away" are collected and sorted. Then the materials are manufactured into new products for which a market exists.

Pickup and Processing. Recyclables are collected in four primary methods: curbside, drop-off centers, buyback centers, and deposit programs.

(501)
Regardless of the collection method, recyclables are sent to a materials recovery facility to be sorted and prepared into marketable commodities for manufacturing.

Manufacturing. Once cleaned and separated, the recyclables are ready to be manufactured into a marketable product. Four major manufacturing sectors accounted for more than half of the recycling activity: paper mills, steel mills, plastic converters, and iron and steel foundries.

The principal obstacle to recycling of plastic is that plastic types must not be mixed, yet it is impossible to tell one type from another by sight or touch. The plastic industry has responded to this problem by developing a series of numbers marked inside triangles on the bottom of containers. Types 1 and 2 are commonly recycled, the others generally are not. Glass is 100 percent recyclable and can be used repeatedly with no loss in quality. Scrap aluminum is readily accepted for recycling, although other metals are rarely accepted.

Other Pollution by Reduction Strategies
In addition to recycling, two other basic strategies can reduce pollution. The amount of waste discharged into the environment can be reduced, or the capacity of the environment to accept discharges can be expanded.

Reducing Discharges. Pollution can be prevented if the amount of waste being discharged into the environment is reduced to a level that the environment can assimilate. The mix of various inputs can be adjusted to produce a higher ratio of product to waste. The amount of waste can also be reduced if the production system produces less of the product —or if production ceases altogether—because of lower consumer demand.

Emissions-trading systems can reduce discharges, especially into the atmosphere. To reduce sulfur dioxide discharges, the United States introduced a market through an amendment to the 1990 Clean Air Act. Power companies can buy and sell allowances to emit sulfur dioxide. The Chicago Climate Exchange opened in 2003 to promote reduction of greenhouse gases.

Increasing Environmental Capacity. The second way to handle pollution is to increase environmental capacity to accept waste discharges. The capacity of air, water, and land to accept waste is not fixed, but varies among places and at different times. A deep, fast-flowing river has a greater capacity to absorb wastewater than a shallow, slow-moving one. Wastewater can be stored when the river level is low and released when the river is high. Similarly, exhaust released into stagnant air irritates, whereas exhaust released in windy conditions is quickly dispersed.

(502)
Environmental capacity can also be increased by transforming the waste so that it is discharged into a resource that has the capacity to assimilate it. For example, a coal-burning power plant discharges gases into the atmosphere, causing air pollution. To reduce air pollution, wet scrubbers are installed to wash particulates from the gas before it is released to the atmosphere. Wet scrubbers capture the particulates in water, which then can be discharged into a stream. If the stream is polluted by the discharge, then the wastewater can be cleaned in a settling basin where the particulates drop out. This transforms the residue into a solid waste for disposal on land.

A Coking Plant: Using All Reduction Strategies
A coking plant provides an example of applying all pollution-reduction strategies—recycling, reducing discharges, and increasing environmental capacity.

Comparing Pollution Reduction Strategies
Relying on an increase in the capacity of the environment to accept discharges is risky. Recent history is filled with examples of wastes discharged in the environment with the belief that they would be dispersed or isolated safely: CFCs in the stratosphere, garbage offshore, and toxic chemicals beneath Love Canal.

(503)
Tall smokestacks . . . were successful at dispersing sulfur over a larger area. But the result of the dispersal was that acid precipitation . . . fell hundreds of kilometers away.

A pollutant such as sulfur dioxide might exist at tolerable levels in the air, but it damages trees when it accumulates in the soil. Reducing discharges into the environment (by either changing the production process or recycling) is usually the preferred alternative.

Key Issue 4. Why Can Resources Be Conserved?
- **Sustainable development**
- **Biodiversity**

Because it is one part natural science and one part social science, geography is especially sensitive to the importance of protecting the natural environment while meeting human needs. "Conservation" is a concept that reflects balance between nature and society. **Conservation** is the sustainable use and management of natural resources such as wildlife, water, air, and Earth deposits to meet human needs, including food, medicine, and recreation.

Conservation differs from **preservation**, which is maintenance of resources in their present condition, with as little human impact as possible. The concept of preservation does not regard nature as a resource for human use. In contrast, conservation is compatible with development but only if natural resources are utilized in a careful rather than a wasteful
manner.

An increasingly important approach to careful utilization of resources is sustainable development, based on promotion of biodiversity.

Sustainable Development
Sustainable development is "development that meets the needs of the present without compromising the ability of future generations to meet their own needs," according to the United Nations.

Sustainability and Economic Growth
The U.N.'s "sustainable development" was defined in the 1987 Brundtland Report. Environmental protection, economic growth, and social equity are linked because economic development aimed at reducing poverty can at the same time threaten the environment. A rising level of economic development generates increased pollution, at least until a country reaches a GDP of about $5,000 per person. Consequently, twentieth-century environmental improvements in the more developed countries of North America and Western Europe are likely to be offset by increased pollution in LDCs during the twenty-first century.

The Brundtland Report was optimistic that environmental protection could be promoted at the same time as economic growth and social equity. In recent years the World Bank and other international development agencies have embraced the concept of sustainable development. Planning for development involves consideration of many more environmental and social issues today than was the case in the past.

(504)
Sustainability's Critics
Some environmentally oriented critics have argued that it is too late to discuss sustainability. The World Wildlife Federation (WWF), for example, claims that the world
surpassed its sustainable level around 1980. Others criticize sustainability from the opposite perspective: human activities have not exceeded Earth's capacity, because resource availability has no maximum. Earth's resources have no absolute limit, because the definition of resources changes drastically and unpredictably over time. Environmental improvements can be achieved through careful assessment of the outer limits of Earth's capacity. Critics and defenders of sustainable development agree that one important recommendation of the UN report has not been implemented: increased international cooperation to reduce the gap between more developed and less developed countries.

Biodiversity
Biological diversity, or biodiversity for short, refers to the variety of species across Earth as a whole or in a specific place. Sustainable development is promoted when biodiversity of a particular place or Earth as a whole is protected.

Biological and Geographic Biodiversity
Species variety can be understood from several perspectives. Geographers are especially concerned with biogeographic diversity, whereas biologists are especially concerned with genetic diversity. Estimates of Earth's total number of species range from 3 to 100 million, with 10 million as a median

"guess," meaning that humans have not yet "discovered," classified, and named most of Earth's species.

For geographers, biodiversity is measurement of the number of species within a specific region or habitat. A community containing a large number of species is said to be species-rich, whereas an area with few species is species-poor.

(505)
Two communities may have the same number of species and the same total population of individuals, yet one may be more diverse than the other, depending on the distribution of the total population among the various species. Strategies to protect genetic diversity have been established on a global scale. Strategies to protect biogeographic diversity vary among countries. Frustrated by the inability to precisely measure environmental impacts, Millennium Ecosystem Assessment has undertaken a multiyear effort to establish systematic data sets.

Biodiversity in the Tropics
Reduction of biodiversity through species extinction is especially important in tropical forests. Three species per hour are extinguished in the tropics, and more than 5,000 species are considered in danger of extinction.

Tropical forests occupy only 7 percent of Earth's land area but contain more than one-half of the world's species, including two-thirds of vascular plant species and one-third of avian species. The principal cause of the high rate of extinction is cutting down forests. Rapid deforestation results from changing economic activities in the tropics, especially a
decline in shifting cultivation (see Chapter 10).

Governments in developing countries support the destruction of rain forests, because they view activities such as selling timber to builders or raising cattle for fast-food restaurants as more effective strategies for promoting economic development than shifting cultivation. Until recently, the World Bank has provided loans to finance development proposals that require clearing forests. The problem with shifting cultivation compared to other forms of agriculture is that it can support only a low level of population in an area without causing environmental damage.

Tropical rain forests are disappearing at the rate of 10 to 20 million hectares (25 to 50 million acres) per year. The amount of Earth's surface allocated to tropical rain forests has been reduced to less than one-half of its original area during the past quarter century. And unless drastic measures are taken, the area will be reduced by another one-fifth within a decade.

Key Terms:

Acid deposition (p.491)
Acid precipitation (p.491)
Active solar energy systems (p.498)
Air pollution (p.490)
Animate power (p.475)
Biochemical oxygen demand (BOD) (p.496)
Biodiversity (p.504)
Biomass fuel (p.475)
Breeder reactor (p.485)
Chlorofluorocarbon (CFC) (p.491)
Conservation (p.503)
Ferrous (p.487)

Fission (p.483)
Fossil fuel (p.476)
Fusion (p.499)
Geothermal energy (p.499)
Greenhouse effect (p.490)
Hydroelectric power (p.499)
Inanimate power (p.475)
Nonferrous (p.487)
Nonrenewable energy (p.476)
Ozone (p.491)
Passive solar energy systems (p.498)